Hermann Minkowski

Space and Time
Minkowski's Papers on Relativity

Translated by Fritz Lewertoff and Vesselin Petkov

Edited by Vesselin Petkov

MINKOWSKI
Institute Press

Hermann Minkowski
22 June 1864 – 12 January 1909

ISBN 978-0-9879871-4-3 (softcover)
ISBN 978-0-9879871-1-2 (ebook)

Minkowski Institute Press
Montreal, Quebec, Canada
minkowskiinstitute.org/mip/

For information on all Minkowski Institute Press publications visit our website at minkowskiinstitute.org/mip/books/

May the hope be fulfilled, through this dissertation, that a wider circle of people become motivated so that participants, who immerse themselves in Minkowski's ideas and the theory of relativity, may each and all contribute their part to promote and spread this theory in accordance with Minkowski's bold dream and that, hence, future generations of mankind will be consciously aware that space and time recede completely to become mere shadows and only the space-time-transformation still stays alive.

Aachen, May 1910
Otto Blumenthal

From the Foreword to H. Minkowski, *Zwei Abhandlungen über die Grundgleichungen der Elektrodynamik* (Teubner, Leipzig 1910)

Preface

This volume contains together for the first time Hermann Minkowski's three papers on relativity written by himself[1]:

- *The Relativity Principle*[2] – lecture given at the meeting of the Göttingen Mathematical Society on November 5, 1907.

- *The Fundamental Equations for Electromagnetic Processes in Moving Bodies*[3] – lecture given at the meeting of the Göttingen Scientific Society on December 21, 1907.

- *Space and Time*[4] – lecture given at the 80th Meeting of the Natural Scientists in Cologne on September 21, 1908.

The three papers were translated by Fritz Lewertoff and myself. Fritz Lewertoff translated *Das Relativitätsprinzip*, which is the first English trans-

---

[1] Almost immediately after Minkowski's sudden and untimely departure M. Born (a student of Minkowski) embarked on decoding the calculations Minkowski left and succeeded in assembling them in a fourth paper (which has never been translated into English and which will be included in a planned volume with Minkowski's physics papers to be published also by the *Minkowski Institute Press*): Eine Ableitung der Grundgleichungen für die elektromagnetischen Vorgänge in bewegten Körpern vom Standpunkte der Elektronentheorie (Aus dem Nachlaß von Hermann Minkowski bearbeitet von Max Born. *Mathematische Annalen* **68** (1910) S. 526-551); reprinted in H. Minkowski, *Zwei Abhandlungen über die Grundgleichungen der Elektrodynamik, mit einem Einführungswort von Otto Blumenthal* (Teubner, Leipzig 1910) S. 58-82, and in *Gesammelte Abhandlungen von Hermann Minkowski*, ed. by D. Hilbert, 2 vols. (Teubner, Leipzig 1911), vol. 2, pp. 405-430.

[2] H. Minkowski, Das Relativitätsprinzip, *Annalen der Physik* **47** (1915) S. 927-938.

[3] H. Minkowski, Die Grundgleichungen für die elektromagnetischen Vorgänge in bewegten Körpern, *Nachrichten der K. Gesellschaft der Wissenschaften zu Göttingen. Mathematisch-physikalische Klasse* (1908) S. 53-111; reprinted in H. Minkowski, *Zwei Abhandlungen über die Grundgleichungen der Elektrodynamik, mit einem Einführungswort von Otto Blumenthal* (Teubner, Leipzig 1910) S. 5-57, and in *Gesammelte Abhandlungen von Hermann Minkowski*, ed. by D. Hilbert, 2 vols. (Teubner, Leipzig 1911), vol. 2, pp. 352-404.

[4] H. Minkowski, Raum und Zeit, *Physikalische Zeitschrift* **10** (1909) S. 104-111; *Jahresbericht der Deutschen Mathematiker-Vereinigung* **18** (1909) S. 75-88; reprinted in *Gesammelte Abhandlungen von Hermann Minkowski*, ed. by D. Hilbert, 2 vols. (Teubner, Leipzig 1911), vol. 2, pp. 431-444, and in H.A. Lorentz, A. Einstein, H. Minkowski, Das Relativitätsprinzip (Teubner, Leipzig 1913) S. 56-68. This lecture also appeared as a separate publication (booklet): H. Minkowski, *Raum und Zeit* (Teubner, Leipzig 1909).

lation, and the Dedication – the last paragraph of Otto Blumenthal's Foreword to H. Minkowski, *Zwei Abhandlungen über die Grundgleichungen der Elektrodynamik* (Teubner, Leipzig 1910). I translated the other two papers.

My initial intention was to retranslate (by making corrections wherever necessary) the only English translation of *Die Grundgleichungen für die elektromagnetischen Vorgänge in bewegten Körpern* done in 1920 by Saha[5], but since I was anyway checking every single sentence and also typesetting the paper in LaTeX I ended up with a virtually new translation.

Raum und Zeit was translated anew. I would like to thank Fritz Lewertoff for his invaluable advice on the translation of three difficult passages and for his patience – our discussions often lasted between one and two hours.

Montreal
July 2012

Vesselin Petkov

---

[5] *The Principle of Relativity: Original Papers by A. Einstein and H. Minkowski*, Translated into English by M.N. Saha and S.N. Bose with a Historical Introduction by P.C. Mahalanobis. (The University of Calcutta, Calcutta 1920). Minkowski's paper *Die Grundgleichungen für die elektromagnetischen Vorgänge in bewegten Körpern* is translated in this book under the name "Principle of Relativity" and the paper's Appendix *Mechanics and the Relativity Postulate* had been extended by the inclusion of the translation of Minkowski's paper *Raum und Zeit*.

Contents

Introduction

The not-fully-appreciated Minkowski

The major reason for the publication of Minkowski's papers on relativity is to correct an injustice which has been ongoing – Minkowski's contributions to modern physics have not been fully and appropriately appreciated. The very fact that so far his papers on relativity have not been published together either in German or English (and even his *Das Relativitätsprinzip* has never been translated into English) is an indication of that.

Since the first publication[1] in April 1908 of Minkowski's mathematical formalism of what he regarded as a theory of an absolute four-dimensional world there have been never stopping attempts to downplay his revolutionary contributions to the modern spacetime physics. Here are several examples:

- Unfortunately, it was Einstein himself (with Jakob Laub) who expressed the first documented[2] reservation towards Minkowski's four-dimensional physics. Einstein and Laub indicated in the first paragraph of their first paper on Minkowski's study *Die Grundgleichungen für die elektromagnetischen Vorgänge in bewegten Körpern* that "In view of the fact that this study makes rather great demands on the reader in its mathematical aspects, we do not consider it superfluous to derive here these important equations in an elementary way, which, is, by the way, essentially in agreement with that of Minkowski"[3]. Ein-

---

[1] H. Minkowski, Die Grundgleichungen für die elektromagnetischen Vorgänge in bewegten Körpern, *Nachrichten der K. Gesellschaft der Wissenschaften zu Göttingen. Mathematisch-physikalische Klasse* (1908) S. 53-111. This is the lecture Minkowski gave at the meeting of the Göttingen Scientific Society on December 21, 1907.

[2] A. Einstein, J. Laub, Über die elektromagnetischen Grundgleichungen für bewegte Körper. *Annalen der Physik* **26** (1908) S. 532-540; Über die im elektromagnetischen Felde auf ruhende Körper ausgeübten ponderomotorischen Kräfte. *Annalen der Physik* **26** (1908) S. 541-550.

[3] *The Collected Papers of Albert Einstein, Volume 2: The Swiss Years: Writings, 1900-*

stein called Minkowski's approach "superfluous learnedness"[4] (über-flüssige Gelehrsamkeit). Also, Sommerfeld's recollection of what Einstein said on one occasion can provide further indication of his initial attitude towards Minkowski's development of the implications of the equivalence of the times of observers in relative motion: "Since the mathematicians have invaded the relativity theory, I do not understand it myself any more"[5].

- Sommerfeld understood and accepted Einstein's special relativity thanks to Minkowski's four-dimensional formulation. That is why it is difficult to explain why he made changes to the original text of Minkowski's lecture *Das Relativitätsprinzip* given at the meeting of the Göttingen Mathematical Society on November 5, 1907, which he prepared for publication in 1915. Sommerfeld's changes were favourable to Einstein as Pyenson[6] observed: "Sommerfeld was unable to resist rewriting Minkowski's judgement of Einstein's formulation of the principle of relativity. He introduced a clause inappropriately praising Einstein for having used the Michelson experiment to demonstrate that the concept of absolute space did not express a property of phenomena. Sommerfeld also suppressed Minkowski's conclusion, where Einstein was portrayed as the clarifier, but by no means as the principal expositor, of the principle of relativity." Giving credit to Einstein for realizing the crucial role of the Michelson experiment is especially unfortunate since Einstein himself stated the opposite: "In my own development, Michelson's result has not had a considerable influence. I even do not remember if I knew of it at all when I wrote my first paper on the subject (1905). The explanation is that I was, for general reasons, firmly convinced that there does not exist absolute motion and my problem was only how this could be reconciled with our knowledge of electrodynamics. One can therefore understand why in my personal

---

1909 (Princeton University Press, Princeton 1989), p. 329.

[4] A. Pais, *Subtle Is the Lord: The Science and the Life of Albert Einstein* (Oxford University Press, Oxford 2005) p. 152.

[5] A. Sommerfeld, To Albert Einstein's Seventieth Birthday. In: *Albert Einstein: Philosopher-Scientist.* P. A. Schilpp, ed., 3rd ed. (Open Court, Illinois 1969) pp. 99-105, p. 102.

[6] L. Pyenson, Hermann Minkowski and Einstein's Special Theory of Relativity, *Archive for History of Exact Sciences* 17 (1977) pp. 71-95, p. 82; see also L. Corry, Hermann Minkowski and the Postulate of Relativity, *Archive for History of Exact Sciences* **51** (1997) p. 273-314, p. 276 and P. L. Galison, Minkowski's Space-Time: From Visual Thinking to the Absolute World, *Historical Studies in the Physical Sciences*, **10** (1979) pp. 85-121, p. 93.

struggle Michelson's experiment played no role, or at least no decisive role."[7] Minkowski's view of the role of Einstein's 1905 paper in clarifying the physical meaning of the Lorentz transformations is expressed at the end of the first part of his 1908 paper *The Fundamental Equations for Electromagnetic Processes in Moving Bodies* (see this volume): "The paper of Einstein which has been cited in the Introduction, has succeeded to some extent in presenting the nature of the transformation from a physical standpoint."

- Despite his initial negative reaction towards Minkowski's four-dimensional physics Einstein relatively quickly realized that his revolutionary theory of gravity would be impossible without the revolutionary contributions of Minkowski. At the beginning of his 1916 paper on general relativity Einstein wrote: "The generalization of the theory of relativity has been facilitated considerably by Minkowski, a mathematician who was the first one to recognize the formal equivalence of space coordinates and the time coordinate, and utilized this in the construction of the theory." This quote is hardly from the new 1997 translation.[8] Quite strangely, the first page of the paper containing the recognition of Minkowski's work had been omitted in the first English translation.[9]

- Many physicists (including relativists) do not appear to have been fully appreciating the depth of Minkowski's four-dimensional physics and his general explanation of relativistic phenomena – "The whole world presents itself as resolved into such worldlines, and I want to say in advance, that in my understanding the laws of physics can find their most complete expression as interrelations between these worldlines" (this volume). In 1960 Synge wrote: "It is to support Minkowski's way of looking at relativity that I find myself pursuing the hard path of the missionary. When, in a relativistic discussion, I try to make things clearer by a space-time diagram, the other participants look at it with polite detachment and, after a pause of embarrassment as if some childish indecency had been exhibited, resume the debate in their own terms"[10]. Now the situation does not appear to be that bad, but it

---

[7] A. Pais, *Subtle Is the Lord: The Science and the Life of Albert Einstein* (Oxford University Press, Oxford 2005) p. 172.

[8] *The Collected Papers of Albert Einstein, Volume 6: The Berlin Years: Writings, 1914-1917* (Princeton University Press, Princeton 1997), p. 146.

[9] H. A. Lorentz et al., *The Principle of Relativity*, translated by W. Perrett and G. B. Jeffery (Methuen 1923; Dover repr., 1952).

[10] J. L. Synge, *Relativity: the general theory* (North-Holland, Amsterdam 1960) p. IX.

is not much better either – everyone can check how many kinematical relativistic effects are explained through spacetime diagrams in recent textbooks on relativity. Given the fact that it is *only* Minkowski's four-dimensional physics that provides *the correct explanations* of the relativistic effects (see below and also the next section), it is difficult to understand the reluctance and sometimes even resistance against explaining the kinematical relativistic effects as manifestations of the four-dimensionality of the world as Minkowski advocated. A possible but disturbing explanation may be an approach that appears to be held by some physicists – that it is merely *a matter of description* whether we will use Einstein's or Minkowski's versions of special relativity. I think such an approach is a sure recipe for a double failure – in genuinely understanding physical phenomena and in making discoveries in physics – because it is certainly not a matter of description whether the world is three- or four-dimensional.

- There have been authors of books on general relativity, spacetime and gravitation, including of recent (21st century) ones, who abundantly use Minkowski's four-dimensional mathematical formalism and spacetime concepts introduced by him, but in a whole book mention his name just once, for example. I prefer not to give any references.

- What is also unfortunate is that some well-known physicists who write papers and books for the general public virtually do not mention Minkowski's contributions and often omit even his name. As a result most who have read about spacetime appear to believe it was introduced by Einstein.

- There have been claims by different authors that Minkowski did not understand Einstein's special relativity. The actual situation had been just the opposite as will be shown in the next section.

Minkowski and Einstein

Let me make it clear right away – it is not my intention at all to try to downplay Einstein's contributions to special relativity. As stated at the beginning of the Introduction the main purpose of this book is to correct an injustice towards Minkowski, and an injustice cannot be corrected by committing another injustice. I hope it would be fair to both Minkowski and Einstein to shed some additional light (based on the historical facts we know now) on what they knew and understood in the period 1905-1908. I think the best approach in such situations is to imagine that they both were alive and would read what is written about them.

Let me start with very brief information about Minkowski's academic background (Einstein's background is well-known) and several facts.

In April 1883 the French Academy granted the Grand Prize in Mathematics jointly to the eighteen year old Hermann Minkowski for his innovative geometric approach to the theory of quadratic forms and to Henry Smith. Thirteen years later, in 1896, Minkowski published his major work in mathematics. *The Geometry of Numbers*[11]

By 1905 Minkowski was already internationally recognized as an exceptional mathematical talent. At that time he became interested in the electron theory and especially in an unresolved issue at the very core of fundamental physics – at the turn of the nineteenth and twentieth century Maxwell's electrodynamics had been interpreted to show that light is an electromagnetic wave, which propagates in a light carrying medium (the luminiferous ether), but its existence was put into question since Michelson's interference experiments failed to detect the Earth's motion in that medium. Minkowski's documented involvement with the electrodynamics of moving bodies began in the summer of 1905 when he and his friend David Hilbert co-directed a seminar in Göttingen on the electron theory. The paper of Minkowski's student – Einstein – on special relativity was not published at that time; *Annalen der Physik* received the paper on June 30, 1905. Poincaré's longer paper "Sur la dynamique de l'électron" was not published either; it appeared in 1906. Also, "Lorentz's 1904 paper (with a form of the transformations now bearing his name) was not on the syllabus."[12]

Minkowski's student Max Born, who attended the seminar in 1905, recalled in 1959 what Minkowski had said during the seminar:[13] "I remember that Minkowski occasionally alluded to the fact that he was engaged with the Lorentz transformations, and that he was on the track of new interrelationships." Again Born wrote in his autobiography about what he had heard from Minkowski after Minkowski's lecture "Space and Time" given on September 21, 1908:[14] "He told me later that it came to him as a great shock

[11]H. Minkowski, *Geometrie der Zahlen* (Teubner, Leipzig 1896).

[12]S. Walter, Minkowski, Mathematicians, and the Mathematical Theory of Relativity, in H. Goenner, J. Renn, J. Ritter, T. Sauer (eds.), *The Expanding Worlds of General Relativity*, Einstein Studies, volume 7, (Birkhäuser, Basel 1999) pp. 45-86, p. 46.

[13]Quoted from T. Damour, "What is missing from Minkowski's "Raum und Zeit" lecture", *Annalen der Physik* **17** No. 9-10 (2008), pp. 619-630, p. 626.

[14]M. Born, *My Life: Recollections of a Nobel Laureate* (Scribner, New York 1978) p.

when Einstein published his paper in which the equivalence of the different local times of observers moving relative to each other were pronounced; for he had reached the same conclusions independently but did not publish them because he wished first to work out the mathematical structure in all its splendour. He never made a priority claim and always gave Einstein his full share in the great discovery."

These facts and especially the results of Minkowski's publications are the best proof that in the period 1905-1908 Minkowski had found a truly revolutionary resolution of the difficult issues surrounding the electrodynamics of moving bodies – that the relativity principle implies, as will be briefly summarized below, that the Universe is a four-dimensional world with time as the fourth dimension. Unfortunately, Minkowski had never indicated exactly when he arrived at that discovery. In any case, it had been sufficiently long before his December 1907 lecture *The Fundamental Equations for Electromagnetic Processes in Moving Bodies* when he presented the fully developed mathematical formalism of the four-dimensional physics introduced by him (a formalism that could not have been created in just several months).

So in the fall of 1907 Minkowski was the only one who had genuine understanding of a number of difficult and unresolved at that time issues:

- The profound *physical meaning of the relativity principle* – that physical phenomena are the same for all inertial observers in relative motion. As a mathematician it may have been easier for Minkowski (than for Einstein) to postulate that the (real) time t of a stationary observer and the abstract mathematical time t', which Lorentz introduced calling it the *local time* of a moving observer, are equivalent and to explore the consequences of such a hypothesis. Unfortunately, we will never know how Minkowski arrived at the idea that t and t' should be treated equally. What appears certain is that his path had been different from Einstein's. The mathematical way of thinking surely had helped Minkowski to realize that if two observers in relative motion have different times they necessarily must have different spaces as well (since space is perpendicular to time), which is impossible in a three-dimensional world, but in a four-dimensional world with time as the fourth dimension. Here is how Minkowski in his own words at his lecture *Space and Time* explained how he had realized the profound *physical meaning of the relativity principle* – that the world is four-dimensional. In the case of two inertial reference frames in relative

---

131.

motion along their x-axes "one can call t' time, but then must necessarily, in connection with this, define space by the manifold of three parameters x', y, z in which the laws of physics would then have exactly the same expressions by means of x', y, z, t' as by means of x, y, z, t. Hereafter we would then have in the world no more *the* space, but an infinite number of spaces analogously as there is an infinite number of planes in three-dimensional space. Three-dimensional geometry becomes a chapter in four-dimensional physics" (this volume). Minkowski suddenly found the answers to many questions in his four-dimensional physics, e.g. the answer to the question of why the relativity principle requires that physical phenomena be the same in all inertial reference frames – this is so because every inertial observer describes the phenomena in *exactly the same way* – in his own reference frame (i.e. in terms of his own space and time) in which he is *at rest*. Also, the answer to the question of the failure of Michelson's experiments to detect the motion of the Earth appears obvious – the Earth is at rest with respect to its space and therefore not only Michelson's but any other experiments would confirm this state of rest. As every observer always measures the velocity of light (and anything else) in his own (rest) space and by using his own time, the velocity of light is the same for all observers.

- Minkowski's realization that the relativity principle implies many times and spaces, which in turn implies that the world is four-dimensional, naturally explained why there is no absolute motion (since there are many spaces, not just one absolute space), and why there is a difference between inertial and accelerated motion (a body moving by inertia is represented by a straight timelike worldline, whereas the worldline of an accelerated body is curved). Minkowski found it necessary to stress that "Especially the concept of *acceleration* acquires a sharply prominent character" (this volume). This sharply prominent character of the acceleration comes from the *absolute* geometric property of the worldline of an accelerated body – the worldline of such a body is curved (deformed); therefore the absoluteness of acceleration merely reflects the absolute fact that the worldline an accelerating body is curved (deformed) and does not imply an absolute space with respect to which the body accelerates.

- Minkowski's four-dimensional physics allowed him not only to *explain* the physical meaning of length contraction, but to realize clearly that, exactly like the relativity principle, that effect is also a manifestation of the four-dimensionality of the world.

- In his four-dimensional physics Minkowski found that pairs of ordinary mechanical quantities are in fact space and time components of four-dimensional vectors and the ordinary electromagnetic quantities are components of new types of four-dimensional structures.

Einstein won the race with his mathematics professor Minkowski (of the existence of which neither of them suspected) and first published his special relativity in 1905 in which he postulated the equivalence of t and t'. The realization of this equivalence took him many years and it came as a result of the persistent analysis of his thought experiment of racing a light beam. This thought experiment became a paradox for Einstein when he studied Maxwell's equations at the Polytechnic Institute in Zurich. In Maxwell's theory the velocity of light is a universal constant $(c = (\mu_0\epsilon_0)^{-1/2})$ which meant for Einstein (due to his trust in "the truth of the Maxwell-Lorentz equations in electrodynamics" and that they "should hold also in the moving frame of reference."[15]) that if he travelled almost at the speed of light (relative, say, to Earth), a beam of light would still move away from him at velocity c, which is in Einstein's own words "in conflict with the rule of addition of velocities we knew of well in mechanics"[16] Later Einstein acknowledged that "the germ of the special relativity theory was already present in that paradox"[17] and explained that his "solution was really for the very concept of time, that is, that time is not absolutely defined but there is an inseparable connection between time and the signal velocity. With this connection, the foregoing extraordinary difficulty could be thoroughly solved. Five weeks after my recognition of this, the present theory of special relativity was completed."[18]

Einstein's realization that inertial observers in relative motion have different times had been accomplished through conceptual analyses *à la* Galileo. The development of this powerful method had later helped Einstein to make one of the greatest discoveries in the intellectual history of our civilization – that gravitational phenomena are not caused by gravitational forces but are a manifestation of the non-Euclidean geometry of spacetime. However, in 1905 Einstein still did not understand fully all implications of his major

---

[15]A. Pais, *Subtle Is the Lord: The Science and the Life of Albert Einstein* (Oxford University Press, Oxford 2005) p. 139

[16]A. Pais, Ibid.

[17]A. Folsing, *Albert Einstein: A Biography* (Penguin Books, New York 1997) p. 166

[18]A. Pais, Ibid.

discovery that t and t' should be treated equally. As a result, at that time and at least in the following several years Einstein did not have complete understanding of the above list of issues which Minkowski clarified in 1907 and 1908. For example, unlike Minkowski Einstein had to postulate the relativity principle without being able to explain its physical meaning. He also simply stated that the luminiferous ether was superfluous without any explanation, that is, he merely postulated that absolute motion does not exist. Einstein did not have the correct understanding of the *physical meaning* of length contraction either since at that time he had not yet fully understood and adopted Minkowski's four-dimensional physics.

One of the indications that Einstein did not fully comprehend the implications of the fact that observers in relative motion have different times is the very name of his theory – the theory of *relativity*. Einstein believed that all uniform motion is relative, whereas Minkowski demonstrated that that relativity is a manifestation of (or implies) an *absolute* four-dimensional world. What is even worse, is that Einstein insisted on *relativity* as the core concept of his theories and called his revolutionary theory of gravitation *the general theory of relativity*, which is a further indication of his slow acceptance of Minkowski's four-dimensional physics. As Synge remarked[19] Minkowski "protested against the use of the word 'relativity' to describe a theory based on an 'absolute' (space-time), and, had he lived to see the general theory of relativity, I believe he would have repeated his protest in even stronger terms."

It is well known that Einstein was "for general reasons, firmly convinced that there does not exist absolute motion"[20] and that Einstein regarded all motion as relative mostly due to Mach. And indeed Einstein kept the term "relativity" in his general theory because he believed that in that theory acceleration should also be treated as relative. In his 1914 paper *The Formal Foundation of the General Theory of Relativity*[21] Einstein repeated and extended Mach's argument for a relative acceleration. This fact alone is sufficient to demonstrate that even in 1914 Einstein had not fully understood Minkowski's spacetime physics.[22] As indicated above Minkowski

---

[19] J. L. Synge, *Relativity: the general theory* (North-Holland, Amsterdam 1960) p. IX.

[20] A. Pais, loc. cit., p. 172.

[21] *The Collected Papers of Albert Einstein, Volume 6: The Berlin Years: Writings, 1914-1917* (Princeton University Press, Princeton 1997) p. 31.

[22] However, later in his life Einstein seems to have fully realized the implications of spacetime not only for physics but for our entire worldview as well (see last section). Regarding Mach, Einstein wrote in 1954: "As a matter of fact, one should no longer speak of Mach's principle at all" (A. Pais, loc. cit., p. 288).

particularly pointed out the prominent character of the concept of acceleration since the acceleration's absoluteness comes from the absolute fact that the worldline of an accelerating body is curved (deformed). It is true that Minkowski's explanation of the absoluteness of acceleration was given for the case of flat spacetime, whereas in 1914 Einstein was completing his theory of general relativity. However, the situation regarding the absoluteness of acceleration is exactly the same in the case of curved spacetime (i.e. in general relativity) – a body moving by inertia is represented by a geodesic worldline (which is the analog of a straight worldline in curved spacetime since it is curved only due to the curvature of spacetime, but is not additionally curved, i.e. it is not deformed), whereas an accelerating body is represented by a deformed (non-geodesic) worldline. Therefore acceleration in both flat and curved spacetime is absolute which demonstrate that Mach's view of relative acceleration is clearly wrong. Here is a concrete example to see why this is so. Mach argued that one could not say anything about the state of motion of a *single* particle in the Universe since he believed that one can talk only about motion *relative* to another body. However, that situation is crystal clear in Minkowski's spacetime physics – the worldline of a single particle in the Universe is either geodesic or deformed, which means that the particle is either moving by inertia or accelerating.

Despite the difficulties Einstein had had with understanding and adopting Minkowski's spacetime physics, the mastering of the method of conceptual analyses involving thought experiments helped him draw all three-dimensional implications of the equivalence of the times of observers in relative motion. For example, the thought experiments led Einstein to the relation between mass and energy $E = mc^2$ which now bears his name although it was discovered before him in the framework of the electron theory.[23]

In view of all these facts it is inexplicable how could anyone say that Minkowski had not understood Einstein's 1905 paper on special relativity. I will give two examples which are even more inexplicable since they come from the authors of two very informative and otherwise excellent papers.

In 1979 Galison[24] wrote: "At this early time (1907) it is clear that Minkowski did not understand the import of Einstein's theory." As we have seen the actual situation had been just the opposite. Galison had in mind

---

[23]When it was initially derived in the electron theory that expression contained the famous factor of 4/3, which was later accounted for; see V. Petkov, *Relativity and the Nature of Spacetime*, 2nd ed. (Springer, Heidelberg 2009) Chap. 9, particularly Sec. 9.3 and the references therein.

[24]P. L. Galison, Minkowski's Space-Time: From Visual Thinking to the Absolute World, Historical Studies in the Physical Sciences, Vol. 10 (1979) pp. 85-121, p. 93.

Minkowski's enthusiasm for arriving at an electromagnetic picture of the world based on his world postulate and the electron theory as suggested by the last paragraph of *Space and Time* (this volume): "The validity without exception of the world postulate is, I would think, the true core of an electromagnetic world view which, as Lorentz found it and Einstein further unveiled it, lies downright and completely exposed before us as clear as daylight." First, not only in 1907 but also in 1908 (when *Space and Time* was presented in Cologne) Minkowski had the same view; moreover his Cologne lecture essentially explained in a non-technical language the main results of his lecture given on December 21, 1907. And I do not see anything wrong with Minkowski's hope for a unified world picture; at that time the other fundamental interactions were unknown, so it was perfectly natural to try to find a unified picture of the world on the basis of what was known. Most important, however is the following. If "Minkowski did not understand the import of Einstein's theory" because he was positively looking at the electron theory, then by exactly the same argument Einstein did not understand the import of his own theory. In January 1909 Einstein wrote[25] "In conclusion, I would also like to point to the importance of the recently published paper by Ph. Frank, which, by taking into account the Lorentz contraction, restores the agreement between Lorentz's treatment, based on the electron theory, and Minkowski's treatment of the electrodynamics of moving bodies. The advantage of the treatment based on the electron theory consists, on the one hand, in providing a graphic interpretation of the field vectors and, on the other hand, in dispensing with the arbitrary assumption that the derivatives of the velocity of matter do not appear in the differential equations." As seen from this quote, in 1909 Einstein viewed "Minkowski's treatment of the electrodynamics of moving bodies" as different from Lorentz' treatment *"based on the electron theory"* and pointed out the "advantage of the treatment based on the electron theory."

Now the prevailing view is that the electron theory was wrong. I am afraid that that is rather a simplistic view. It is now clear what in the electron theory was undoubtedly wrong – e.g. the electron is not a small charged sphere. A completely wrong theory cannot make a number of correct predictions – e.g. the electron theory predicted that the electron mass increases as the electron's velocity increases *before* the theory of relativity, yielding the correct velocity dependence, and that the relation between en-

---

[25] A. Einstein, Comment on the paper of D. Mirimanoff "On the Fundamental Equations..." *Annalen der Physik* **28** (1909) pp. 885-888. In: *The Collected Papers of Albert Einstein, Volume 2: The Swiss Years: Writings, 1900-1909* (Princeton University Press, Princeton 1989), p. 356.

ergy and mass is $E = mc^2$. That is why it is maybe more appropriate to say that today "the state of the classical electron theory reminds one of a house under construction that was abandoned by its workmen upon receiving news of an approaching plague. The plague in this case, of course, was quantum theory. As a result, classical electron theory stands with many interesting unsolved or partially solved problems."[26]

Unfortunately, exactly a hundred years after Minkowski's lecture *Space and Time* Damour[27] wrote: "First, I would like (after many others...) to stress that Minkowski probably did not really comprehend the conceptual novelty of Einstein's June 1905 paper on Special Relativity, and especially the results therein concerning *time*. Indeed, in his Cologne lecture Minkowski says that, while Einstein "deposed [time] from its high seat", "neither Einstein nor Lorentz made any attack on the concept of space..." However, this was precisely one of the key new insights of Einstein, namely the *relativity of simultaneity!*"

Now, *thanks to Minkowski*, we know that relativity of simultaneity does imply many spaces since *a space constitutes a class of simultaneous events –* two observers in relative motion have different classes of simultaneous events and therefore different spaces and vice versa (as Minkowski discovered two observers in relative motion have different spaces and therefore different classes of simultaneous events). However, in 1905 Einstein was *totally unaware* of this. He had been occupied with the idea of time and how to *measure* times and distances. Even a quick look at how Einstein arrived at the idea of relativity of simultaneity in his 1905 paper shows that he did that in an *operational way* – by analyzing the *procedure* of synchronizing distant clocks through light signals; relativity of simultaneity follows immediately from the fact that the velocity of light is c for all observers. That is why Einstein *himself* had never claimed that he had realized that observers in relative motion have different spaces. On the contrary, as indicated above *three years after* his 1905 paper (in May 1908) he reacted negatively towards the introduced by Minkowski absolute four-dimensional world and therefore negatively towards the very idea of many spaces since *it was the idea of many spaces* that led Minkowski to the absolute four-dimensional world. As we saw above Minkowski's geometrical approach helped him to realize first that as observers in relative motion have different times they necessarily must have different spaces as well, and then he had probably immediately

---

[26]P. Pearle, Classical Electron Models. In: *Electromagnetism: Paths to Research*, ed. by D. Teplitz (Plenum Press, New York 1982) pp. 211-295, p. 213.

[27]T. Damour, "What is missing from Minkowski's "Raum und Zeit" lecture", *Annalen der Physik.* **17**, No. 9-10, (2008) pp. 619-630, p. 627.

seen that many spaces imply an absolute four-dimensional world.

As unfounded as the statement above (that Einstein had discovered that observers in relative motion have different spaces), is another statement in Damour's article:

> In addition, when Minkowski introduces the (geometrically motivated) concept of proper time, he does not seem to fully grasp its physical meaning. However, this is the second key new insight brought in by Einstein concerning time, namely the fact (explicitly discussed by Einstein) that, when comparing a moving clock to one remaining at rest (and marking the corresponding 'rest' coordinate time t), the moving clock will mark (upon being reconvened with the sedentary clock) the time
>
> $$\tau = \int dt \sqrt{1 - v^2/c^2}$$
>
> i.e. Minkowski's proper time. It seems that Minkowski was not aware of this.

Minkowski was certainly aware of this expression *without the integral* (there is no integral in Einstein's paper as Damour admits but in a footnote) – on October 9, 1907 he wrote to Einstein to request a copy of his 1905 paper.[28] Damour's suggestion that Minkowski might have misread the paper – "This is another example of a scientist misreading a paper which he knew, however, to be central to his research topic!"[29] – seems virtually impossible since "Minkowski had written to Einstein asking for a reprint of his 1905 paper, in order to study it in his joint seminar with Hilbert"[30] (could Minkowski have misread a *key* paper that had been *studied* at the seminar he co-directed with Hilbert?).

What is most important, however, is that, like the above issue of many spaces, Damour again seems to read more in Einstein's 1905 paper. Einstein had completed that paper only five weeks after he had realized the equivalence of the times of observers in relative motion and had been still struggling with its consequences. By contrast, Minkowski seems to have had

---

[28]S. Walter, Minkowski, Mathematicians, and the Mathematical Theory of Relativity, in H. Goenner, J. Renn, J. Ritter, T. Sauer (eds.), *The Expanding Worlds of General Relativity*, Einstein Studies, volume 7, (Birkhäuser, Basel 1999) pp. 45-86, p. 47.

[29]T. Damour, loc. cit., p. 627.

[30]L. Corry, Hermann Minkowski and the Postulate of Relativity, *Archive for History of Exact Sciences* **51** (1997) p. 273-314, p. 276.

more than two years to explore those consequences – Minkowski appears ho have realized independently the equivalence of the times of observers in relative motion almost certainly as late as the summer of 1905.[31] The best proof that Minkowski fully understood the physical meaning of proper time (which is quite natural given that this concept was introduced by himself) is the fact that the modern introduction and definition of proper time is *identical* to that of Minkowski. Only an in-depth and complete understanding of the new concepts of space and time and their union made their introduction and definition so precise that they remained unchanged more than a hundred years later. As this should be self-evident since it was Minkowski who thoroughly developed these new concepts it is inexplicable why not only did Damour make the above claim but found it necessary to repeat it: "Minkowski did not fully grasp the physical meaning of what he was doing."[32]

Minkowski's understanding of the physical meaning of time and spacetime had been so deep that with the introduction of proper time he essentially demonstrated that an observer should use *two* times in his rest frame – *proper* and *coordinate* times (τ and t) – which provided the correct physical treatment of time (i) in accelerated reference frames in special relativity, and later (ii) in general relativity. Minkowski did not call the time t coordinate time, but the presence of the *two* times in the *same* reference frame is obvious from the way he defined proper time (this volume):

$$d\tau = \frac{1}{c}\sqrt{c^2 dt^2 - dx^2 - dy^2 - dz^2}.$$

The expression $c^2 dt^2 - dx^2 - dy^2 - dz^2$ is the interval (the spacetime distance) ds^2 (in a reference frame) between the two infinitesimally close events on the worldline of a particle; the length of the worldline between these events is the proper time $d\tau$. If the particle's worldline is straight, which means that the particle moves with constant velocity, in its inertial reference frames proper and coordinate times coincide. However, if the particle accelerates,

---

[31]There are two indications of that which cannot be merely ignored. First, Born's recollection quoted in the first section that Minkowski had been shocked when Einstein's paper appeared in 1905; there is no reason whatsoever to suspect that Born would invent such a recollection (moreover, he had another recollection, as indicated also in the first section, which supports it). Second, what is far more important, however, is the full-blown four-dimensional formalism Minkowski reported on December 21, 1907 and the depth of his understanding of the electrodynamics of moving bodies and the absolute four-dimensional world; such a revolution in both physics and mathematics could not have been possible if he had merely developed others' ideas.

[32]T. Damour, loc. cit., p. 627.

its worldline is curved and an observer in the particle's accelerating frame should use both proper and coordinate times.

If Damour had insisted on keeping in his paper the repeated unfortunate expression "did not fully grasp the physical meaning of what he was doing," he should have used it for Einstein's understanding of the physical meaning of the time (in the case discussed by Damour) which Minkowski later called proper time (but that would have been equally unfair since as indicated above Einstein completed his 1905 paper only five weeks after his profound insight that the times of observers in relative motion should be treated equally). In the above calculation quoted by Damour, Einstein determined the time of a clock in *circular* motion: "If there are two synchronous clocks in A, and one of them is moved along a closed curve with constant velocity[33] until it has returned to A, which takes, say, t sec, then this clock will lag on its arrival at A $\frac{1}{2}t(v/V)^2$ sec behind the clock that has not been moved."[34] Einstein arrived at this result by using the Lorentz transformation of the times of two inertial clocks in relative motion, which generally deals with *coordinate* time. As coordinate and proper time coincide in inertial reference frames (moving with constant velocity) no misunderstanding is likely. But in an accelerating reference frame coordinate and proper time do not coincide. When Einstein compared the times of the *accelerating* clock (moving along the closed curve) and the stationary clock he used what was later called the proper time of the accelerating clock without having any idea that that time is a *second* time in the reference frame of the accelerating clock, which is *different* from the coordinate time (Minkowski introduced the concept of proper time more than two years later).

Damour further wrote[35] that Minkowski "had (seemingly) not fully grasped the striking result of Einstein that proper time along any polygonal (or curved) time-like line between two points in spacetime is smaller than the proper time along the straight line joining the two points. If he had realized it clearly, he would have commented that this is just the opposite of the usual triangular inequality." First, the wording of "the striking result of Einstein that proper time..." is inappropriate – it is well known and indicated above that in 1905 Einstein could not have had any idea of what

---

[33] Even in the new translation of Einstein's 1905 paper the German word *Geschwindigkeit* has been again erroneously translated in this sentence as *velocity*. Obviously, the velocity of the clock along a closed curve is not constant; what is constant is the clock's speed.

[34] A. Einstein, On the electrodynamics of moving bodies, *The Collected Papers of Albert Einstein, Volume 2: The Swiss Years: Writings, 1900-1909* (Princeton University Press, Princeton 1989), p. 153.

[35] T. Damour, loc. cit., p. 629.

proper time is. Second, as Minkowski defined proper time as a length along a timelike worldline he knew perfectly what proper time is, and it is indeed a valid question why he did not define the triangle inequality in spacetime as well.

I think the most probable explanation is that since he had been completely occupied with developing the spacetime physics and its four-dimensional mathematical formalism his first priority had been (as seen from his three papers) the electrodynamics of moving bodies. The work on the kinematical consequences of the absolute four-dimensional world (e.g. the special role of acceleration stressed by Minkowski) had been scheduled for later as Minkowski clearly alluded to such a plan: "The whole world presents itself as resolved into such worldlines, and I want to say in advance, that in my understanding the laws of physics can find their most complete expression as interrelations between these worldlines" (this volume). The triangle inequality is clearly such an interrelation between worldlines.

To expect more from someone who had already done so much for such a short period of time, and who would have indisputably done even more, if he had not been taken away from us when he was at the peak of his intellectual strength, is very unfair.

It is important to stress that after his initial hostile attitude towards Minkowski's spacetime physics Einstein gradually adopted it since it was essential for his general relativity. In 1946 in his Autobiography Einstein summarized Minkowski's main contribution:[36]

> Minkowski's important contribution to the theory lies in the following: Before Minkowski's investigation it was necessary to carry out a Lorentz-transformation on a law in order to test its invariance under such transformations; he, on the other hand, succeeded in introducing a formalism such that the mathematical form of the law itself guarantees its invariance under Lorentz-transformations. By creating a four-dimensional tensor-calculus he achieved the same thing for the four-dimensional space which the ordinary vector-calculus achieves for the three spatial dimensions. He also showed that the Lorentz-transformation (apart from a different algebraic sign due to the special character of time) is nothing but a rotation of the coordinate system in the four-dimensional space.

---

[36]A. Einstein, "Autobiographical notes." In: *Albert Einstein: Philosopher-Scientist.* Paul A. Schilpp, ed., 3rd ed. (Open Court, Illinois 1969) pp. 1-94, p. 59.

As seen from his estimation of Minkowski's contribution Einstein did not explicitly credit Minkowski for demonstrating that the relativity postulate and length contraction imply an absolute four-dimensional world; we will return to this point in the last section. On the other hand, Einstein credited Minkowski for showing that the Lorentz transformations are rotations in spacetime, whereas it was Poincaré who first published that result in 1906.[37]

Let me stress it one more time – Einstein's achievements speak for themselves, so no one can downplay his contributions. I think Minkowski's four-dimensional (spacetime) physics and Einstein's discovery that gravity is a manifestation of the spacetime curvature will forever remain as the two greatest intellectual achievements. The approaches of Minkowski and Einstein are distinctly different, but they both proved to be so extraordinarily productive that should become integral parts of the way of thinking of any scientist who works on the front line of research in any field. Minkowski's and Einstein's proven but not fully studied approaches form the core of a research strategy that is being developed and will be employed at a new research institute – *Institute for Foundational Studies 'Hermann Minkowski'* (http://minkowskiinstitute.org/).

In addition, I have a personal reason not to even think of downplaying Einstein's contributions. I have always admired him for the way he arrived at his two theories – by employing and extending Galileo's way of doing physics through conceptual analyses and thought experiments. Moreover, my own way of thinking about physical phenomena was consciously formed by studying the methods of great physicists which led them to groundbreaking discoveries, particularly those of Galileo and Einstein; much later I discovered and started to appreciate thoroughly Minkowski's approach to physics.

Also, I fully share Einstein's firm position that quantum mechanics does not provide a complete description of the quantum world in a sense that it does not contain a model of the quantum object *itself*. I believe a theory that describes only the *state* of something, not the something itself, is intrinsically incomplete. As now no one can seriously question the probabilistic nature of quantum phenomena it appears easily tempting to state that Einstein's intuition that God does not play dice was wrong. I think such a temptation will remain baseless until we understand what the quantum object is.

Leaving aside the issue of whether God would care about a human's opinion on how he should behave, just imagine the following (very prob-

---

[37]H. Poincaré, Sur la dynamique de l'électron. *Rendiconti del Circolo matematico Rendiconti del Circolo di Palermo* **21** (1906) pp.129-176.

able in my view) development in quantum physics, which may reveal an unanticipated meaning of Einstein's intuition. As Galileo's and Einstein's conceptual analyses (which proved to be physics at its best) are now almost explicitly regarded as old-fashioned (no leading physics journal would publish a paper containing a deep conceptual analysis of an open question), it is not surprising that the so called quantum paradoxes remained unresolved almost a century after the advent of quantum mechanics.

Despite Feynman's desperate appeal to regard Nature as absurd[38] the history of science teaches us that all apparent paradoxes are caused by some implicit assumptions. A consistent conceptual analysis of only one of those quantum mechanical paradoxes – say, the famous double-slit experiment, discussed by Feynman – almost immediately identifies an implicit assumption[39] – we have been taking for granted that quantum objects exist *continuously* in time although there has been nothing either in the experimental evidence or in the theory that compels us to do so. Just imagine – a fundamental *continuity* (continuous existence in time) at the heart of quantum physics. And no wonder that such an implicit assumption leads to a paradox – an electron, for example, which is always registered as a pointlike entity and which exists continuously in time, is a classical particle (i.e. a worldline in spacetime) that cannot go simultaneously through both slits in the double-slit experiment to form an interference pattern.

However, if we abandon the implicit assumption and replace it explicitly with its alternative – discontinuous existence in time – the paradox disappears. Then an electron is, in the ordinary three-dimensional language, an ensemble[40] of constituents which appear-disappear $\sim 10^{20}$ times per second (the Compton frequency). Such a quantum object can pass simultaneously through *all* slits at its disposal.

In Minkowski's four-dimensional language (trying to extract more from his treasure), an electron is not a worldline but a "disintegrated" worldline whose worldpoints are scattered all over the spacetime region where the electron wavefunction is different from zero. Such a model of the quantum object and quantum phenomena in general provides a surprising insight

---

[38]Feynman wrote: "The theory of quantum electrodynamics describes Nature as absurd from the point of view of common sense. And it agrees fully with experiment. So I hope you can accept Nature as She is – absurd." R. P. Feynman, *QED: The Strange Theory of Light and Matter* (Princeton University Press, Princeton 1985) p. 10.

[39]V. Petkov, *Relativity and the Nature of Spacetime*, 2nd ed. (Springer, Heidelberg 2009) Chap. 10.

[40]A. H. Anastassov, Self-Contained Phase-Space Formulation of Quantum Mechanics as Statistics of Virtual Particles, *Annuaire de l'Universite de Sofia "St. Kliment Ohridski"*, *Faculte de Physique* **81** (1993) pp. 135-163.

into the physical meaning of probabilistic phenomena in spacetime – an electron is a *probabilistic* distribution of worldpoints which is *forever given* in spacetime.

Had Minkowski lived longer he might have described such a spacetime picture by the mystical expression "predetermined probabilistic phenomena." And, I guess, Einstein would be also satisfied – God would not play dice since a probabilistic distribution in spacetime exists eternally there.

Minkowski and Poincaré

This section was the most difficult to write since I have not found any clue of how Minkowski would have explained the obvious fact – that Poincaré was not mentioned in his Cologne lecture *Space and Time*. Minkowski was certainly aware of Poincaré's paper *Sur la dynamique de l'électron* published in 1906 (but received by *Rendiconti del Circolo matematico Rendiconti del Circolo di Palermo* on July 23, 1905) since he quoted it in his previous lectures given in November and December 1907. In his paper Poincaré first published the important result that the Lorentz transformations had a geometric interpretation as rotations in what he seemed to have regarded as an *abstract* four-dimensional space with time as the fourth dimension.[41]

Here are two attempts to explain Minkowski's omission to mention Poincaré's paper in his Cologne lecture.

> In the absence of any clear indication why Minkowski left Poincaré out of his lecture, a speculation or two on his motivation may be entertained. If Minkowski had chosen to include some mention of Poincaré's work, his own contribution may have appeared derivative. Also, Poincaré's modification of Lorentz's theory of electrons constituted yet another example of the cooperative role played by the mathematician in the elaboration of physical theory. Poincaré's "more mathematical" study of Lorentz's electron theory demonstrated the mathematician's dependence upon the insights of the theoretical physicist, and as such, it did little to establish the independence of the physical and mathematical paths to the Lorentz group. The metatheoretical goal of establishing the essentially mathematical nature of the principle of relativity was no doubt more easily attained by neglecting Poincaré's

---

[41]H. Poincaré, Sur la dynamique de l'électron, *Rendiconti del Circolo matematico Rendiconti del Circolo di Palermo* **21** (1906) pp. 129-176.

elaboration of this principle.[42]

> My conjecture is that Minkowski, helped by his background reading of some of the works of Lorentz and Poincaré (which, however, *did not include* their most recent contributions of 1904-1905...) had discovered by himself, in the summer of 1905 (without knowing about the 1905 papers of Poincaré) the fact that Lorentz transformations preserve the quadratic form $-c^2t^2 + x^2(+y^2 + z^2)$. If that reconstruction is correct, he must have been all the more eager, when he later realized that he had been preceded by Poincaré, to find reasons for downplaying Poincaré's work.[43]

I think one should also ask why in 1946 in his Autobiography[44] (as quoted in Section 2) Einstein wrote that Minkowski "showed that the Lorentz-transformation [...] is nothing but a rotation of the coordinate system in the four-dimensional space." It seems Einstein was either unaware in 1946 (which is highly unlikely) of the fact that it was Poincaré who first published that result, or he knew (perhaps from Born) that Minkowski independently had made the same discovery.

Another interesting fact is that not someone else but a famous *French* physicist credited Minkowski for the discovery of spacetime. In 1924 Louis de Broglie wrote in his doctoral thesis *Recherches sur la théorie des quanta*:[45] "Minkowski showed first that one obtains a simple geometric representation of the relationships between space and time introduced by Einstein by

---

[42]S. Walter, Minkowski, Mathematicians, and the Mathematical Theory of Relativity, in H. Goenner, J. Renn, J. Ritter, T. Sauer (eds.), *The Expanding Worlds of General Relativity*, Einstein Studies, volume 7, (Birkhäuser, Basel 1999) pp. 45-86, p. 58.

[43]T. Damour, What is missing from Minkowski's "Raum und Zeit" lecture, *Annalen der Physik*. **17**, No. 9-10, (2008) pp. 619-630, p. 626.

[44]A. Einstein, "Autobiographical notes." In: *Albert Einstein: Philosopher-Scientist*. Paul A. Schilpp, ed., 3rd ed. (Open Court, Illinois 1969) pp. 1-94, p. 59.

[45]"Minkowski a montré le premier qu'on obtenait une représentation géométrique simple des relations de l'espace et du temps introduites par Einstein en considérant une multiplicité euclidienne à 4 dimensions dite Univers ou Espace-temps," Louis de Broglie, *Recherches sur la théorie des quanta*, Réédition du texte de 1924. (Masson, Paris 1963), p. 27. Strangely, the word "appears" (which is clearly not in the original French text) had been inserted into the sentence translated into English by Kracklauer: "Minkowski appears to have been first to obtain a simple geometric representation of the relationships introduced by Einstein between space and time consisting of a Euclidian 4-dimensional space-time," Louis-Victor de Broglie, *On the Theory of Quanta*, translated by A. F. Kracklauer (2004); available at the website of *Annales de la Fondation Louis de Broglie* (http://aflb.ensmp.fr/LDB-oeuvres/De_Broglie_Kracklauer.htm).

considering an Euclidean manifold of 4 dimensions called Universe or space-time." Another contemporary French physicist – Thibault Damour (quoted above) – also thinks that "the replacement of the separate categories of space and time with the new physical category of space-time is [...] more properly attributed to Hermann Minkowski and not to Poincaré."[46]

Probably we will never learn why Minkowski did not quote Poincaré in his lecture *Space and Time* in 1908. However, a similar question applies to Poincaré himself: "In the lecture Poincaré delivered in Göttingen on the new mechanics in April 1909, he did not see fit to mention the names of Minkowski and Einstein."[47] Poincaré could have used the fact that his lecture was only around three months after Minkowski's death to credit Minkowski for fully developing the four-dimensional physics based on the idea of spacetime which Poincaré first published.

I think the discovery of spacetime is a doubly sad story. First, unlike Minkowski, Poincaré seems to have seen nothing revolutionary in the idea of a mathematical four-dimensional space as Damour remarked[48] – "although the first discovery of the mathematical structure of the space-time of special relativity is due to Poincaré's great article of July 1905, Poincaré (in contrast to Minkowski) had never believed that this structure could really be important for physics. This appears clearly in the final passage that Poincaré wrote on the question some months before his death":

> Everything happens as if time were a fourth dimension of space, and as if four-dimensional space resulting from the combination of ordinary space and of time could rotate not only around an axis of ordinary space in such a way that time were not altered, but around any axis whatever...
>
> What shall be our position in view of these new conceptions? Shall we be obliged to modify our conclusions? Certainly not; we had adopted a convention because it seemed convenient and we had said that nothing could constrain us to abandon it. Today some physicists want to adopt a new convention. It is not that they are constrained to do so; they consider this new convention more convenient; that is all. And those who are not of this

---

[46]T. Damour, *Once Upon Einstein*, Translated by E. Novak (A. K. Peters, Wellesley 2006) p. 49.

[47]S. Walter, Minkowski, Mathematicians, and the Mathematical Theory of Relativity, in H. Goenner, J. Renn, J. Ritter, T. Sauer (eds.), *The Expanding Worlds of General Relativity*, Einstein Studies, volume 7, (Birkhäuser, Basel 1999) pp. 45-86, p. 57.

[48]T. Damour, loc. cit., p. 51.

opinion can legitimately retain the old one in order not to disturb their old habits. I believe, just between us, that this is what they shall do for a long time to come.[49]

Poincaré believed that our physical theories are only *convenient descriptions* of the world and therefore it is really a matter of *convenience* and *our choice* which theory we would use. As Damour stressed it, it was "the sterility of Poincaré's scientific philosophy: complete and utter "conventionality" [...] which stopped him from taking seriously, and developing as a physicist, the space-time structure which he was the first to discover."[50]

What makes Poincaré's failure to comprehend the profound physical meaning of the relativity principle and the geometric interpretation of the Lorentz transformations especially sad is that it is perhaps the most cruel example in the history of physics of how an inadequate philosophical position can prevent a scientist, even as great as Poincaré, from making a discovery. However, this sad example can serve a noble purpose. Science students and young scientists can study it and learn from it because scientists often think that they do not need any philosophical position for their research:

> Scientists sometimes deceive themselves into thinking that philosophical ideas are only, at best, decorations or parasitic commentaries on the hard, objective triumphs of science, and that they themselves are immune to the confusions that philosophers devote their lives to dissolving. But there is no such thing as philosophy-free science; there is only science whose philosophical baggage is taken on board without examination.[51]

---

[49]H. Poincaré, *Mathematics and Science: Last Essays (Dernières Pensées)*, Translated by J.W. Bolduc (Dover, New York 1963) pp. 23-24. Poincaré even appeared to have thought that the spacetime convention would not be advantageous: "It quite seems, indeed, that it would be possible to translate our physics into the language of geometry of four dimensions. Attempting such a translation would be giving oneself a great deal of trouble for little profit, and I will content myself with mentioning Hertz's mechanics, in which something of the kind may be seen. Yet, it seems that the translation would always be less simple than the text, and that it would never lose the appearance of a translation, for the language of three dimensions seems the best suited to the description of our world, even though that description may be made, in case of necessity, in another idiom." H. Poincaré, *Science and Method*, In: *The Value of Science: Essential Writings of Henri Poincaré* (Modern Library, New York 2001) p. 438.

[50]T. Damour, loc. cit., p. 52.

[51]D. C. Dennett, *Darwin's Dangerous Idea: Evolution and the Meanings of Life* (Simon and Schuster, New York 1996) p. 21.

Second, it seems virtually certain that Minkowski independently arrived at two important results – (i) the equivalence of the times of observers in relative motion and (ii) the fact that the Lorentz transformations preserve the quadratic form $c^2t^2 - x^2 - y^2 - z^2$ and can therefore be regarded geometrically as rotation in a four-dimensional space with time as the fourth dimension. But these results were first published by Einstein and Poincaré, respectively. As indicated in Section 2 the best proof that Minkowski, helped by his extraordinary geometrical imagination, had made these discoveries independently of Einstein and Poincaré, is the introduced by him four-dimensional (spacetime) physics with a fully developed mathematical formalism and his deep understanding of the new worldview and its implications. Born's recollections given in Section 2 only confirm what follows from a careful study of Minkowski's results.

Minkowski and gravitation

On January 12, 1909 only several months after his Cologne lecture *Space and Time* at the age of 44 Minkowski tragically and untimely departed from this strange world (as Einstein would call it later). We will never know how physics would have developed had he lived longer.

What seems undeniable is that the discovery of the true cause of gravitation – the non-Euclidean geometry of spacetime – would have been different from what actually happened. On the one hand, Einstein's way of thinking based on conceptual analyses and thought experiments now seems to be the only way powerful enough to decode the unimaginable nature of gravitation. However, on the other hand, after Minkowski had written the three papers on relativity included here, he (had he lived longer) and his friend David Hilbert might have formed an unbeatable team in theoretical physics and might have discovered general relativity (surely under another name) before Einstein.

As there is no way to reconstruct what might have happened in the period 1909-1915 I will outline here what steps had been logically available to Minkowski on the basis of his results. Then I will briefly discuss whether their implications would lead towards the modern theory of gravitation – Einstein's general relativity.

In 1907 (most probably in November) Einstein had already been well ahead of Minkowski when he made a gigantic step towards the new theory of gravity:[52]

---

[52]Quoted from: A. Pais, *Subtle Is the Lord: The Science and the Life of Albert Einstein*

In the first row of this photograph (probably taken around 1905) are Minkowski
(left) David Hilbert's wife, Käthe, and David Hilbert. Source: D. E. Rowe, A Look
Back at Hermann Minkowski's Cologne Lecture "Raum und Zeit," *The Mathematical Intelligencer*, Volume 31, Number 2 (2009), pp. 27-39.

> I was sitting in a chair in the patent office at Bern when all of a
> sudden a thought occurred to me: "If a person falls freely he will
> not feel his own weight." I was startled. This simple thought
> made a deep impression on me. It impelled me toward a theory
> of gravitation.

Einstein had been so impressed by this insight that he called it the
"happiest thought" of his life.[53] And indeed this is a crucial point – at that
time Einstein had been the only human who realized that no gravitational
force acted on a falling body. Then he struggled eight years to come up with
a theory – his general relativity – according to which gravity is not a force
but a manifestation of the curvature of spacetime.

Here I will stress particularly the core of general relativity which reflects
Einstein's "happiest thought" – the geodesic hypothesis according to which
a falling particle is not subject to a gravitational force. In other words,
the geodesic hypothesis in general relativity assumes that the worldline of
a free particle is a timelike *geodesic* in spacetime. The geodesic hypothesis

(Oxford University Press, Oxford 2005) p. 179.

[53] A. Pais, Ibid.

is regarded as "a natural generalization of Newton's first law,"[54] that is, "a mere extension of Galileo's law of inertia to curved spacetime."[55] This means that *in general relativity a particle, whose worldline is geodesic, is a free particle which moves by inertia.*

The geodesic hypothesis has been *confirmed* by the experimental fact that particles falling towards the Earth's surface *offer no resistance to their fall* – a falling accelerometer, for example, reads zero resistance (i.e. zero acceleration; the observed *apparent* acceleration of the accelerometer is caused by the spacetime curvature caused by the Earth). The experimental fact that particles do not resist their fall (i.e. their apparent acceleration) means that they move by inertia and therefore no gravitational force is causing their fall. It should be emphasized that a gravitational force would be required to accelerate particles downwards *only if* the particles *resisted* their acceleration, because *only then* a gravitational force would be needed to *overcome* that resistance.

Let us now imagine how Minkowski would have approached the issue of gravitation. By analogy with Maxwell's electrodynamics he had already modified Newton's gravitational theory in order that the speed of gravity be equal to that of light c (Poincaré also proposed such a modification in his 1906 paper on the dynamics of the electron). Now, thanks to the genius of Einstein, we know that electromagnetism is fundamentally different from gravitation – electromagnetic phenomena are caused by electromagnetic *forces*, whereas gravitational phenomena are manifestation of the non-Euclidean *geometry of spacetime* which means that there are no gravitational forces in Nature.

The natural question is whether Minkowski would have found any reasons to revise his modified version of Newton's theory of gravity. Perhaps many physicists would say 'highly unlikely.' And they might be right. But looking at what Minkowski had achieved for so short a period of time, I think his genius should never be underestimated (even because that would constitute a contradiction in terms). Let us see what logical options Minkowski had after his third lecture *Space and Time*.

Minkowski had been aware of two relevant facts – (i) the motion of particles with constant velocity cannot be detected experimentally since the particles move non-resistantly, i.e. by inertia (in other words, an experiment always detects the *lack of resistance* of an inertial particle, and in this

---

[54] J. L. Synge, *Relativity: the general theory.* (Nord-Holand, Amsterdam 1960) p. 110.

[55] W. Rindler, *Relativity: Special, General, and Cosmological* (Oxford University Press, Oxford 2001) p. 178.

sense inertial motion is absolute or frame-independent), and (ii) the accelerated motion of a particle can be discovered experimentally since the particle
resists its acceleration (so accelerated motion is also absolute in this sense
and therefore frame-independent).

The accelerated motion had already been causing problems after the
publication of Einstein's special relativity in 1905 since it appeared that the
experimental detection of accelerated motion provided experimental support for the absolute space – if a particle's acceleration is *absolute* (since
it is measurable), then such an acceleration is with respect to the absolute
space, which contradicts both Einstein's special relativity and particularly
Minkowski's interpretation of the relativity principle according to which observers in relative motion have different times and *spaces* (whereas an absolute space implies a *single* space).

However, Minkowski had not been concerned about such an apparent
contradiction at all. He provided rigorous criteria for inertial and accelerated motion[56] – a free particle, which moves by inertia, is a straight timelike
worldline in Minkowski spacetime, whereas the timelike worldline of an accelerating particle is clearly different – it is *curved* (i.e. *deformed*). That
is why Minkowski wrote at the beginning of Section III of *Space and Time*:
"Especially the concept of *acceleration* acquires a sharply prominent character."

These criteria show that in spacetime the absoluteness of inertial (non-
resistant) and accelerated (resistant) motion become more understandable
– the straightness of a timelike worldline (representing inertial motion) and
the curvature or rather the *deformation* of a timelike worldline (representing
accelerated motion) are absolute (frame-independent) properties of worldlines. These absolute properties of worldlines (straightness and deformation)
correspond to the absoluteness (frame-independence) of inertial and accelerated motion in terms of experimental detection – it is an experimental
fact that a particle moving by inertia offers no resistance to its uniform motion, and it is an experimental fact that an accelerating particle resists its
acceleration.

Then, as indicated in Section 2, it becomes evident that absolute acceleration is a mere manifestation of the *deformation* of the worldline of an
accelerating particle and *does not imply some absolute space with respect to
which the particle accelerates*. Exactly in the same way, absolute inertial

---

[56]In the beginning of Section II of his paper *Space and Time* (this volume) Minkowski
wrote: "a straight line inclined to the *t*-axis corresponds to a uniformly moving substantial
point, a somewhat curved worldline corresponds to a non-uniformly moving substantial
point."

motion reflects the straightness of the worldline of an inertial particle and does not imply some absolute space with respect to which the particle moves with constant velocity.

Perhaps Minkowski knew all this well. What is more important, however, is that he certainly knew that an accelerating particle is represented by a curved (deformed) worldline. Then he might have realized that inertia – the *resistance* a particle offers to its acceleration – could be regarded as arising from a four-dimensional stress[57] in the deformed worldline, or rather worldtube, of an accelerating particle. Certainly, Minkowski would have been enormously pleased with such a discovery because inertia would have turned out to be another manifestation of the four-dimensionality of the absolute world since only a *real* four-dimensional worldtube could resist its deformation (by analogy with an ordinary deformed rod which resists its deformation). Of course, the question of whether or not Minkowski could have noticed this surprising four-dimensional explanation of the origin of inertia will forever remain unanswerable; but that explanation of inertia follows *logically* from the fact that an accelerating particle is a *deformed* worldtube and therefore would have been a legitimate logical option for Minkowski, especially given the fact that all his contributions to mathematics and physics demonstrated his innovative ability to explore the deep logical structure of what he studied.

We saw that Minkowski's spacetime criteria for inertial and accelerated motion spectacularly resolved the old (since Newton) question of the meaning of absolute acceleration – the acceleration of a particle is absolute not because it accelerates with respect to an absolute space, but because the particleŠs worldline is curved (deformed) which is an absolute geometric property. Then by asking the obvious question "What is the link between the two absolute properties of an accelerating particle – the absolute geometric property (the deformation of its worldline) and the absolute physical property reflected in the fact that an accelerating particle resists its acceleration?" we are led to the surprising insight about the origin of inertia – the resistance a particle offers to its acceleration is in fact the static resistance in the deformed worldline of the accelerating particle.

To see even better the enormous potential of Minkowski's criteria for inertial and accelerated motion let us imagine two scenarios.

First, imagine that Minkowski or someone else who had had profound understanding of Minkowski's spacetime physics had read Galileo's works.

---

[57]V. Petkov, *Relativity and the Nature of Spacetime*, 2nd ed. (Springer, Heidelberg 2009) Chap. 9.

That would have played the role of Einstein's "happiest thought" because Galileo came close to the conclusion that a falling body does not resist its fall:[58]

> But if you tie the hemp to the stone and allow them to fall freely from some height, do you believe that the hemp will press down upon the stone and thus accelerate its motion or do you think the motion will be retarded by a partial upward pressure? One always feels the pressure upon his shoulders when he prevents the motion of a load resting upon him; but if one descends just as rapidly as the load would fall how can it gravitate or press upon him? Do you not see that this would be the same as trying to strike a man with a lance when he is running away from you with a speed which is equal to, or even greater, than that with which you are following him? You must therefore conclude that, during free and natural fall, the small stone does not press upon the larger and consequently does not increase its weight as it does when at rest.

Then the path to the idea that gravitational phenomena are manifestations of the curvature of spacetime would have been open – the experimental fact that a falling particle accelerates (which means that its worldtube is curved), but offers no resistance to its acceleration (which means that its worldtube is not deformed) can be explained only if the worldtube of a falling particle is *both curved and not deformed*, which is impossible in the flat Minkowski spacetime where a curved worldtube is always deformed. Such a worldtube can exist only in a non-Euclidean spacetime whose geodesics are naturally curved due to the spacetime curvature, but are not deformed.

Second, imagine that after his *Space and Time* lecture Minkowski found a very challenging mathematical problem and did not compete with Einstein for the creation of the modern theory of gravitation. But when Einstein linked gravitation with the geometry of spacetime Minkowski regretted his change of research interests and started to study intensely general relativity and its implications.

As a mathematician he would be appalled by what he saw as confusing of physics and geometry:

- The new theory of gravitation demonstrates that gravitational physics

---

[58]Galileo, *Dialogues Concerning Two Sciences*. In: S. Hawking (ed.), *On The Shoulders Of Giants*, (Running Press, Philadelphia 2002) pp. 399-626, p. 447

is in fact geometry of curved spacetime; no general relativity of anything can be found there.

- How could physicists say that in the framework of general relativity *itself* gravitational phenomena are caused by gravitational *interaction*? According to what general relativity itself tells us *gravity is not a physical interaction* since by the geodesic hypothesis particles falling towards a planet and planets orbiting the Sun *all move by inertia* and inertia by its very nature presupposes *no interaction*. The mass of the Sun, for example, curves spacetime *no matter whether or not there are other planets in its vicinity*, and the planets move by inertia while orbiting the Sun (the correct expression is: the planets' worldlines are geodesics which represent inertial motion).

- How could physicists talk about gravitational energy in the framework of general relativity? There is no gravitational field and no gravitational force; the gravitational field is at best a *geometric* not a physical field, and as such it does not possess any energy. Moreover, the mathematical formalism of general relativity *itself* refuses to yield a proper (tensorial) expression for gravitational energy and momentum.

I guess some physicists might be tempted to declare that such questions are obvious nonsense. For instance, they might say that the decrease of the orbital period of a binary pulsar system, notably the system PSR 1913+16 discovered by Hulse and Taylor in 1974, provided indirect experimental evidence for the existence of gravitational energy that is carried away by the gravitational waves emitted by the neutrons stars in the system.

It may sound heretical, but the assumption that the orbital motion of the neutron stars in the PSR 1913+16 system loses energy by emission of gravitational waves *contradicts general relativity*, particularly the geodesic hypothesis and the experimental evidence which confirmed it. The reason is that by the geodesic hypothesis the neutron stars, whose worldlines are geodesics (the neutron stars in the PSR 1913+16 system had been modeled by Taylor and Hulse "as a pair of orbiting point masses" which means that they are exact geodesics) *move by inertia without losing energy* since the very essence of inertial motion is motion *without any loss of energy*.

Therefore no energy is carried away by the gravitational waves emitted by the binary pulsar system. For this reason the experimental fact of the decay of the orbital motion of PSR 1913+16 (the shrinking of the stars' orbits) does not constitute evidence for the existence of gravitational energy. That fact may most probably be explained in terms of tidal friction as suggested in 1976 as an alternative to the explanation given by Hulse and Taylor.

A detailed critical examination of the "confusing of physics and geometry" (as Minkowski might have called it) is part of an analysis of the nature of inertia and gravitation by explicitly following Minkowski's approach, which is reported in V. Petkov, *Inertia and Gravitation* (Minkowski Institute Press, Montreal 2012), to appear in 2012.

Minkowski and the reality of spacetime

Since 1908 there has been no consensus on the reality of the absolute four-dimensional world no matter whether it is the flat Minkowski spacetime or a curved spacetime since both spacetimes represent a *four-dimensional* world with time *wholly* given as the fourth dimension. What makes this issue truly unique in the history of science is that for over a hundred years not only has it remained an unresolved one, but for some it has been even a non-issue, whereas Minkowski had already provided the necessary evidence for the reality of spacetime in 1907 and 1908. He had fully realized the profound physical meaning of the relativity principle (reflecting the existing experimental evidence) – *the impossibility to discover absolute motion experimentally unequivocally implies that observers in relative motion have different times and spaces, which in turn implies that what exists is an absolute four-dimensional world.*

Apparently Minkowski had realized the entire depth and grandness of the new view of the absolute four-dimensional world imposed on us by the experimental evidence. A draft of his Cologne lecture *Space and Time* reveals that he appears to have tried to tone down his excitement in the announcement of the unseen revolution in our understanding of the world. As the draft shows Minkowski's initial intention had been to describe the impact of the new world view in more detail – he had written that the essence of the new views of space and time "is mightily revolutionary, to such an extent that when they are completely accepted, as I expect they will be, it will be disdained to still speak about the ways in which we have tried to understand space and time."[59] In the final version of the lecture Minkowski had reduced this sentence about the new views of space and time to just "Their tendency is radical."

---

[59]See: P. L. Galison, Minkowski's Space-Time: From Visual Thinking to the Absolute World, *Historical Studies in the Physical Sciences*, **10** (1979) pp. 85-121, p. 98.

Given this rather restrained (compared to the draft version) announcement of the successful decoding of the physical meaning of the relativity principle – that the world is four-dimensional – it is surprising that Damour referred to that announcement as "the somewhat theatrical tone of Cologne's non-technical exposé."[60] The tone of the Cologne lecture could look theatrical only to someone who does not see the major issue in it in the way Minkowski saw it. This seems to be precisely the case since Damour apparently regards Minkowski's unification of space and time into an absolute four-dimensional world as nothing more than a mathematical abstraction:[61]

> Though Minkowski certainly went much farther than Poincaré in taking seriously the 4-dimensional geometry as a new basis for a physico-mathematical representation of reality, it does not seem that he went, philosophically and existentially, as far as really considering 'the flow of time' as an illusory shadow. By contrast, let us recall that the old Einstein apparently did take seriously, at the existential level, the idea that 'time' was an illusory shadow, and that the essence of (experienced) reality was timeless.

Minkowski's paper does not contain anything that even resembles a hint of what Damour wrote – that "it does not seem that he went, philosophically and existentially, as far as really considering 'the flow of time' as an illusory shadow." On the contrary, the whole paper and even its "theatrical tone" (in Damour's own words) unambiguously demonstrates that Minkowski consciously announced a major discovery about the world, not a discovery of a mathematical abstraction (moreover Minkowski was fully aware that that mathematical abstraction was already published by Poincaré two years before Minkowski's Cologne lecture).

It is particularly disturbing when especially relativists do not regard spacetime as representing a *real* four-dimensional world and still hold the *unscientific*[62] view that time flows. Such an opinion of spacetime as nothing

---

[60] T. Damour, "What is missing from Minkowski's "Raum und Zeit" lecture", *Annalen der Physik.* **17**, No. 9-10, (2008) pp. 619-630, p. 620.

[61] T. Damour, loc. cit., p. 626.

[62] This everyday view is unscientific since there is *no scientific evidence* whatsoever for the *sole* existence of the present moment, which is the central element of the concept of time flow (what is sufficient for the issue of the reality of spacetime is that there is no physical evidence for the existence of time flow). If the flow of time were a feature of the physical world (not of the image of the world in our mind), physics would have discovered it by now.

more than a mathematical space was openly defended by another physicist, Mermin, in a recent article *What's bad about this habit* in the May 2009 issue of *Physics Today* where he argued that "It is a bad habit of physicists to take their most successful abstractions to be real properties of our world." [63] He gave the issue of the reality of spacetime as an example – "spacetime is an abstract four-dimensional mathematical continuum" – and pointed out that it is "a bad habit to reify the spacetime continuum". Mermin specifically stressed that spacetime does not represent a real four-dimensional world: "The device of spacetime has been so powerful that we often reify that abstract bookkeeping structure, saying that we inhabit a world that is such a four- (or, for some of us, ten-) dimensional continuum."

I think the proper understanding of Minkowski's spacetime *physics* (which requires more effort than learning its four-dimensional formalism) is crucial not only for deep understanding of modern physics, but more importantly such understanding is a necessary condition for making discoveries in the twenty-first century physics.

The best proof that the *experimental evidence* against the existence of absolute motion (reflected in the relativity postulate) implies that the Universe is an absolute four-dimensional world is contained in Minkowski's paper itself. As discussed in Section 2 Minkowski first realized the important hidden message in the *experimental fact* that physical phenomena are the same in all inertial reference frames (which Einstein merely stated in the relativity postulate without *explaining* it) – physical phenomena are the same in all inertial reference frames because every inertial observer has his own space and time[64] and therefore describes the phenomena in his reference frame (i.e. in his own space and time) in which he is at rest. For example, the Earth is at rest with respect to its space and therefore all experiments confirm this state of rest. Due to his excellent geometrical imagination Minkowski appears to have immediately realized that many spaces are possible in a four-dimensional world. In this way he managed to decode the physical meaning of the experimental fact that absolute motion cannot be discovered – that fact implies that the Universe is an absolute four-dimensional world in which space and time are inseparably amalgamated; *only in such a world one can talk about many spaces and many times*. Minkowski noted that "I think the word *relativity postulate* used for the requirement of invariance under the group G_c is very feeble. Since the meaning of the postulate is

---

[63]N. D. Mermin, What's Bad About This Habit? *Physics Today* 2009, p. 8.

[64]As we saw in Section 2 Minkowski showed that the equivalence of the times of observers in relative motion (which is necessary to explain why absolute motion cannot be detected) means that the observers have not only different times but different spaces as well.

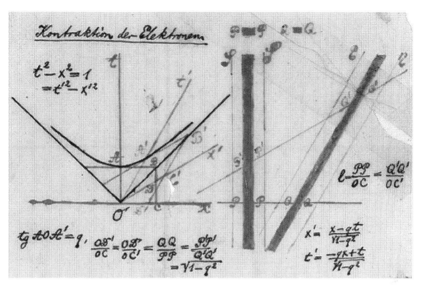

The transparency which Minkowski used at his lecture in Cologne on September 21, 1908. It shows Fig. 1 in his paper (this volume). Source: Cover of *The Mathematical Intelligencer*, Volume 31, Number 2 (2009).

that through the phenomena only the four-dimensional world in space and time is given, but the projection in space and in time can still be made with certain freedom, I want to give this affirmation rather the name *the postulate of the absolute world*" (this volume).

To see why Minkowski's absolute four-dimensional world adequately represents the *dimensionality* of the real world, assume the opposite – that the real world is three-dimensional and time really flows (as our everyday experience so convincingly appears to suggest). Then there would exist just *one* space, which as such would be *absolute* (i.e. it would be the *same* for all observers since only a *single* space would exist). This would imply that absolute motion should exist and therefore there would be no relativity principle.

Another example of why special relativity (as we now call the physics of flat spacetime) would be *impossible in a three-dimensional world* is contained in Minkowski's four-dimensional explanation of the physical meaning of length contraction, which is shown in the above figure (displaying the transparency Minkowski used in 1908). Consider only the vertical (red) strip which represents a body at rest with respect to an observer. The proper length of the body is the cross section PP of the observer's space, represented by the horizontal (red) line, and the body's strip. The relativistically contracted length of the body measured by an observer in relative motion with respect to the body is the cross section $P'P'$ of the moving observer's space, represented by the inclined (green) line, and the body's strip (on the transparency $P'P'$ appears longer than PP because the two-dimensional pseudo-Euclidean spacetime is represented on the two-dimensional Euclidean surface of the page).

To see that no length contraction would be possible in a three-dimensional world,[65] assume that the world is indeed three-dimensional. This would mean that all objects are also three-dimensional. Therefore the four-dimensional vertical strip of the body would not represent anything real in the world and would be merely an abstract geometrical construction. Then, obviously, the cross sections PP and $P'P'$ would coincide and there would be no length contraction since the observers in relative motion would measure the *same* three-dimensional body which has just *one* length $PP \equiv P'P'$.

The impossibility of length contraction in a three-dimensional world also follows even without looking at the spacetime diagram: it follows from the definition of a three-dimensional body – all its parts which exist *simultaneously* at a given moment; when the two observers in relative motion measure the length of the body, they measure *two different three-dimensional bodies* since the observers have *different* sets of simultaneous events, i.e. different sets of simultaneously existing parts of the body (which means two different three-dimensional bodies). If the world and the physical bodies were three-dimensional, then the observers in relative motion would measure the *same* three-dimensional body (i.e. the same set of simultaneously existing parts of the body), which means that (i) they would have a *common set of simultaneous events* in contradiction with relativity (simultaneity would be absolute), and (ii) they would measure the *same* length of the body, again in contradiction with relativity.

The same line of reasoning demonstrates that no relativity of simultaneity, no time dilation, and no twin paradox effect would be possible in a three-dimensional world.[66]

As I gave examples of how some physicists do not fully appreciate the depth of Minkowski's discovery that the physical world is four-dimensional, it will be fair to stress that there have been many physicists (I would like to

---

[65]A visual representation of Minkowski's explanation of length contraction is given in V. Petkov, *Spacetime and Reality: Facing the Ultimate Judge*, Sect. 3 (`http://philsci-archive.pitt.edu/9181/`). I think this representation most convincingly demonstrates that length contraction is impossible in a three-dimensional world.

[66]V. Petkov, *Relativity and the Nature of Spacetime*, 2nd ed. (Springer, Heidelberg 2009) Chap. 5.

think the majority) who have demonstrated in written form their brilliant understanding of what the dimensionality of the world is. Here are several examples.

A. Einstein, *Relativity: The Special and General Theory* (Routledge, London 2001) p. 152:

> It appears therefore more natural to think of physical reality as a four-dimensional existence, instead of, as hitherto, the *evolution* of a three-dimensional existence.

A. S. Eddington, *Space, Time and Gravitation: An Outline of the General Relativity Theory* (Cambridge University Press, Cambridge 1920), p. 51:

> In a perfectly determinate scheme the past and future may be regarded as lying mapped out – as much available to present exploration as the distant parts of space. Events do not happen; they are just there, and we come across them.

A. S. Eddington, *Space, Time and Gravitation: An Outline of the General Relativity Theory* (Cambridge University Press, Cambridge 1920), p. 56:

> However successful the theory of a four-dimensional world may be, it is difficult to ignore a voice inside us which whispers: "At the back of your mind, you know that a fourth dimension is all nonsense." I fancy that that voice must often have had a busy time in the past history of physics. What nonsense to say that this solid table on which I am writing is a collection of electrons moving with prodigious speeds in empty spaces, which relatively to electronic dimensions are as wide as the spaces between the planets in the solar system! What nonsense to say that the thin air is trying to crush my body with a load of 14 lbs to the square inch! What nonsense that the star cluster which I see through the telescope obviously there now, is a glimpse into a past age 50 000 years ago! Let us not be beguiled by this voice. It is discredited.

H. Weyl, *Philosophy of Mathematics and Natural Science* (Princeton University Press, Princeton 2009) p. 116:

The objective world simply *is*, it does not *happen*. Only to the gaze of my consciousness, crawling upward along the life line of my body, does a section of this world come to life as a fleeting image in space which continuously changes in time.

H. Weyl, *Mind and Nature: Selected Writings on Philosophy, Mathematics, and Physics* (Princeton University Press, Princeton 2009) p. 135:

The objective world merely exists, it does not happen; as a whole it has no history. Only before the eye of the consciousness climbing up in the world line of my body, a section of this world "comes to life" and moves past it as a spatial image engaged in temporal transformation.

R. Geroch, *General relativity from A to B* (University of Chicago, Chicago 1978) pp. 20-21:

There is no dynamics within space-time itself: nothing ever moves therein; nothing happens; nothing changes [...] one does not think of particles as 'moving through' space-time, or as 'following along' their world-lines. Rather, particles are just 'in' space-time, once and for all, and the world-line represents, all at once, the complete life history of the particle.

In a real four-dimensional world there is no time flow since *all moments of time have equal existence* as they all form the fourth dimension (which like the other three dimensions is *entirely given*), whereas the very essence of time flow is that *only one moment of time exists* which constantly changes. But it is a well known fact that there does not exist any physical evidence whatsoever that only the present moment exists. On the contrary, all relativistic experimental evidence confirms Minkowski's view that all moments of time have equal existence due to their belonging to the entirely given time dimension. So the old Einstein was wise[67] to take seriously the absolute four-dimensional world and the idea that the flow of time was merely "a stubbornly persistent illusion" as evident from his letter of condolences to the widow of his longtime friend Besso:[68]

---

[67]I think it is this context that is the right and fair one for using the word 'old' especially if it refers to such a scientist and person as Einstein.

[68]Quoted from: Michele Besso, From Wikipedia, the free encyclopedia (`http://en.wikipedia.org/wiki/Michele_Besso`). Besso left this world on 15 March 1955; Einstein followed him on 18 April 1955.

> Now Besso has departed from this strange world a little ahead of me. That means nothing. People like us, who believe in physics, know that the distinction between past, present and future is only a stubbornly persistent illusion.

Minkowski succeeded in demonstrating how the power of mathematical thinking applied to unresolved physical problems can free us from such illusions and can reveal the existence of a reality that is difficult to comprehend at once. Galison masterfully summarized the essence of Minkowski's discovery by pointing out that in his lectures *The Relativity Principle* and *Space and Time* "the idea is the same: beyond the divisions of time and space which are imposed on our experience, there lies a higher reality, changeless, and independent of observer."[69]

I think there are still physicists and philosophers who have been effectively refusing to face the implications of a real four-dimensional world due to the huge challenges they pose. But trying to squeeze Nature into our preset and deceivingly comfortable views of the world should not be an option for anyone in the 21st century.

Montreal
July 2012

Vesselin Petkov

---

[69]P. L. Galison, Minkowski's Space-Time: From Visual Thinking to the Absolute World, *Historical Studies in the Physical Sciences,* **10** (1979) pp. 85-121, p. 98.

The Relativity Principle

With regard to the electromagnetic light theory, it appears that recently a complete transformation that changes our thinking about space and time wants to take place, and to become acquainted with such thinking must in any case be of quite particular interest for a mathematician. Also, the mathematician is particularly well predisposed to absorb these new intuitions because this involves becoming acclimatized and to conceptualize things anew, that is, it involves a process that the mathematician has practiced for the longest time. Although the physicists must now partially and newly invent these concepts whereby they must clear for themselves and with great effort a jungle like path through an unexplored territory, while quite nearby there already exists an excellent highway of the mathematicians' which comfortably leads onwards. After all, the new attempts, if they in fact interpret the phenomena correctly, would present almost the greatest triumph ever that the application of mathematics has brought about as of today. What is being dealt with here is, expressed as foreshortened as possible – I will present a more explicit account later – that the world in space and time in a certain sense is a four dimensional, non-Euclidean manifold. Apparently, as to the fame of mathematicians and the astonishment of the rest of mankind, the mathematicians within their pure fantasies opened up a huge territory, although these ideal craftsmen never had any such intentions, which one day would contain the most complete real existence.

The principle of relativity, to which I want to draw your attention today, has been invented as a means to find an explanation why every experimental attempt which would show that the earth moves relative to a luminiferous aether must necessarily fail. Experiments, which rely on magnitudes of the order of quotients which take the speed of the earth in the solar system over the speed of light as a basis for observation, have shown, so far, that it is impossible to determine the direction of the earth's motion through experiments that take place on the surface of the earth. This is so because circumstances are such that a comparison of two clocks placed at a distance from each other at two separate points must be made whereby signals must necessarily travel forth and back again between these clocks. Moreover, *A. Michelson* performed in 1881 an experiment (which in 1887, together with *Morley*, was repeated on a larger scale) which took into account detection of a second order magnitude in the above mentioned quotients, nevertheless, the result turned out negative just the same. In order to explain this negative result as well, *H. A. Lorentz* (1892) and independently *Fitz Gerald* (1893) formulated the hypothesis that on account of the earth's motion a quite determinable contraction of matter occurs parallel to the earth's motion. From this highly peculiar sounding hypothesis finally evolved the postulate of relativity in a form that particularly suited the mathematician's way of understanding. Credits for the general principle's development are shared by *Einstein, Poincaré*, and *Planck*. I will talk in more detail about their work a little later.

Now I will finally get to the actual subject under discussion and, in order to maintain clarity, I shall divide what follows into four subject headings, namely: 1. Electricity, 2. Matter, 3. Dynamics, and 4. Gravitation.

1. First of all, what is being dealt with here explicitly is a purely mathematical relationship of a certain formal character of those differential equations which Lorentz takes to be fundamental to his electron theory and, moreover, that these equations regulate the functioning of the electromagnetic field in pure aether in the same way as they apply to an infinite space that is filled with electricity. These fundamental equations, although they do not pertain to a right angled coordinate system of space of which they stand quite independently, contain nevertheless a certain additional symmetry which cannot be expressed through our familiar way of writing. Foremost, although this by the way did not happen with regard to the above named authors, not even with Poincaré, I will render here an expression of the above mentioned symmetry, whereby in fact, if I may say so, the form of the equations becomes quite transparent. To start with let x, y, z designate fixed right angled coordinates in space, in the aether, and t designates time. Furthermore, the follow up discussion involves the quadratic expression $x^2 + y^2 + z^2 - c^2t^2$ whereby c denotes the propagation speed of light in empty space. A unit of time may be chosen in such a way that $c = 1$, in other words, as $1/3.10^{10}$ s whereby a unit of length becomes 1 cm. Now one may write x_1, x_2, x_3, instead of x, y, z and furthermore x_4 may replace *it*. In that way, with regard to the following, x_4 stands, of course, for a purely imaginary magnitude. The above quadratic expression may now be transformed into

$$x_1^2 + x_2^2 + x_3^2 + x_4^2$$

which means that from now on we will be dealing with constructions in a four dimensional manifold of x_1, x_2, x_3, x_4. This also indicates that, at any point in time, the entire electromagnetic state in space can now be described through the behaviour of a single four dimensional vector. The components of this vector can be expressed as $\psi_1, \psi_2, \psi_3, \psi_4$ whereby again ψ_1, ψ_2, ψ_3 are real and ψ_4 is purely imaginary. In accord with the descriptions which Abraham gives us in his electromagnetic theory of radiation, $\mathfrak{U}_x, \mathfrak{U}_y, \mathfrak{U}_z$ are the components of the electromagnetic vector potentials $\mathfrak{U}$, whereby $\psi_4 = i\Phi$ and Φ is the scalar electromagnetic potential. This vector (ψ), on account of its dependence upon x_1, x_2, x_3, x_4, must now fulfill the following condition:

$$\frac{\partial \psi_1}{\partial x_1} + \frac{\partial \psi_2}{\partial x_2} + \frac{\partial \psi_3}{\partial x_3} + \frac{\partial \psi_4}{\partial x_4} = 0 \tag{1}$$

The differential expression to the left may be called $\mathrm{Div}(\psi)$. In addition a second four dimensional vector gains validity, namely $(\varrho) = \varrho_1, \varrho_2, \varrho_3, \varrho_4$. Therefore, $\varrho_4 = i\varrho$ whereby ϱ indicates the density of electricity per unit volume, and $\varrho_1, \varrho_2, \varrho_3$ are the components of the spatial vectors $\varrho \mathfrak{b}$ according to Abraham, whereby $\mathfrak{b}$ indicates the velocity of the convective motion of electricity. Finally, the abbreviation $\square$ can be used with the differential expression

$$\frac{\partial^2}{\partial x_1^2} + \frac{\partial^2}{\partial x_2^2} + \frac{\partial^2}{\partial x_3^2} + \frac{\partial^2}{\partial x_4^2}$$

which allows that we take the combinations

$$\frac{\partial \psi_k}{\partial x_j} - \frac{\partial \psi_j}{\partial x_k}$$

so that we may designate the latter by ψ_{jk} whereby, in general, $\psi_{jk} = -\psi_{kj}$ and with the same subscript twice $\psi_{jj} = 0$. This now renders the following differential equations valid:

$$\square \psi_j = -\varrho_j \qquad (j = 1, 2, 3, 4) \tag{2}$$

and from the previously given equation (1) obviously follows $\mathrm{Div}(\varrho) = 0$, which is the equation of continuity of electricity. Finally,

$$\psi_{23}, \psi_{31}, \psi_{12}; \qquad \psi_{14}, \psi_{24}, \psi_{34}$$

can be identified with

$$\mathfrak{H}_x, \mathfrak{H}_y, \mathfrak{H}_z; \qquad -i\mathfrak{E}_x, -i\mathfrak{E}_y, -i\mathfrak{E}_z,$$

whereby $\mathfrak{E}_x, \mathfrak{E}_y, \mathfrak{E}_z$ the electrical components and $\mathfrak{H}_x, \mathfrak{H}_y, \mathfrak{H}_z$ the magnetic components describe field strength. These certainly quite understandably and easily arrived at formulas which are the total expression of the Lorentzian foundational fundamental equations, contain naturally in a special sense, if the vector $(\varrho) = 0$ is applied, Maxwell's equations for the electromagnetic processes in pure aether. Now the usual somewhat artful looking expressions for the interference with charges by ponderomotive forces of the field become quite transparent. Let X, Y, Z be the components of this force which should be calculated per unit volume (not at all per unit charge), and besides these magnitudes one must also take a fourth magnitude into consideration, namely, the work done by this force per second which gives us,

$$A = X\mathfrak{b}_x + Y\mathfrak{b}_y + Z\mathfrak{b}_z;$$

so that X, Y, Z, iA are the components of a four dimensional vector X_1, X_2, X_3, X_4, whereby in general we may write

$$X_j = \varrho_1\psi_{j1} + \varrho_2\psi_{j2} + \varrho_3\psi_{j3} + \varrho_4\psi_{j4}$$

so that now, during the motion of electricity, the following expression will always hold:

$$X_1 \mathrm{d}x_1 + X_2 \mathrm{d}x_2 + X_3 \mathrm{d}x_3 + X_4 \mathrm{d}x_4 = 0.$$

Within these equations, with reference to how I have transformed them in my writing, lies a perfectly obvious fact which hereafter is tied to the relativity principle. That is, if instead of x, y, z, t one establishes through a perfectly acceptable linear transformation a new set of primed coordinates x', y', z', t', then the expression $x_1^2 + x_2^2 + x_3^2 + x_4^2$ remains, so to say, invariant; and if one transforms the vector $\psi_1, \psi_2, \psi_3, \psi_4$ in the same way as x_1, x_2, x_3, x_4 were transformed, then the entire system of the above equations also pertains to the corresponding primed system which incorporates the same equations. In that sense, without having to claim a new law which would lead us astray to something that has not been previously contained, we may proclaim it trivial that the fundamental equations of the electron theory allow for the orthogonal transformations of the four dimensional space.

2. We will now turn to a further contemplation of matter. Here we must deal, on the one hand, with electrodynamics and then, on the other hand, with mechanics. To do this, we adopt a point of view under which we are not as yet totally familiar with the prevailing physical laws. Perhaps one day a reduction to a pure doctrine of electricity will become possible, but as for now, particularly in light of *Michelson*'s experiment, it has been shown that, as Einstein so succinctly expresses this, the concept of an absolute state of rest entails no properties that correspond to phenomena. And this fact can easily be explained if we assume that also the equations of the electrodynamics of matter are in any case of such a kind, that they likewise remain invariant under that group which *Poincaré* named after *Lorentz*. Here the relativity principle takes up its place as a truly new physical law, that is, the relativity principle advances a demand for equations of phenomena which have yet to be established. To what conclusions this postulate leads will now be discussed in detail. For the contemplation of matter we must first of all focus on a new vector as a function of x, y, z, t, the visible velocity of mater at any point. Let w_x, w_y, w_z be the components of the velocity at one position of matter, whereby w is the magnitude of the velocity, thus a vector in space corresponds to real motion and not, however, to the case of rest. I will now consider instead a four dimensional vector; I will take the important result in advance, namely, that velocities of matter equal or greater than the velocity of light proves itself to be an absurdity, that therefore $w < 1$ must always hold true. Now I posit that

$$w_x, w_y, w_z, i\sqrt{1 - w^2}$$

$$= w_1, w_2, w_3, \quad w_4,$$

so that w_1, w_2, w_3, w_4 is always a point upon the surface

$$w_1^2 + w_2^2 + w_3^2 + w_4^2 = -1$$

Or, if you want, upon

$$t^2 - x^2 - y^2 - z^2 = 1, \tag{3}$$

and represents at the same time the four dimensional vector from the origin to this point; and this corresponds also to the velocity of zero, that is, something like a real vector represents a state of rest. The non-Euclidean geometry, of which I already spoke in a preliminary sense, now evolves for these velocity vectors. The transformations, of which I spoke just above, become the real transformations of this four dimensional hyperboloid (3) with regard to conjugated diameters, whereby it becomes perfectly clear that one can begin with insert, as t-diameter, the light ray at any predetermined point of this surface; here we are dealing with transformations under which not only spatial coordinates are transformed, but, at the same time and under a certain connection therewith, time is also transformed. In particular, as we have just seen, the velocity of any single point of matter can therefore be transformed to a state of rest, and when this happened, what remains thereafter by choosing the other three conjugated diameters of (3) is nothing other than the freedom of a single orthogonal transformation of the spatial coordinates. After this explanation it should at once be clear, that the invariance of physical laws with regard to Lorentz' group can be expressed accordingly, as I have already done above, namely, that the absolute state of rest entails no properties of phenomena. Although I will talk about mechanics only afterwards, I will insert here, for a clearer understanding of the ideas presented, how Galileo's law of inertia stands in relation to the postulate of relativity. To this end, we will not at once designate the velocity of light by 1, instead, we will stay for now with the general symbol c for the velocity of light. In that case, with respect to a velocity w in the direction of the x-axis, the transformation, through which this velocity is reduced to a state of rest, becomes:

$$x' = \frac{c(x - wt)}{\sqrt{c^2 - w^2}}, \quad y' = y, \quad z' = z, \quad t' = \frac{c^2 t - wx}{c\sqrt{c^2 - w^2}}.$$

Now we see clearly, with respect to a limit under which $c = \infty$, that the above equations can be expressed as:

$$x' = x - wt, \quad y' = y, \quad z' = z, \quad t' = t,$$

that is, new right angled coordinates are simply inserted with respect to an axis system which, with respect to the initial right angled coordinates, moves uniformly with the translational motion of the initial coordinate system. Hereby, and in accordance with the law of inertia, the laws of mechanics should remain unchanged in their expressions. Accordingly now, the law of inertia signifies an invariance of mechanics for the transformations of the expression $x^2 + y^2 + z^2 - c^2t^2$ whereby $c = \infty$, that is, the law of inertia carries the same meaning as does the postulate of relativity for $c = \infty$.

Now we shall take a closer look at how electrodynamics results foundationally from the principle of relativity. Hereby Lorentz' initial attempts are to be seen as particularly necessary. Besides the velocity vector we have to take into consideration two additional formations of matter; here I present the results again in such a way that the invariance of the Lorentz group becomes evident. To start with, I introduce a four dimensional vector, namely,

$$(\sigma) = \sigma_1, \sigma_2, \sigma_3, \sigma_4,$$

and I shall call this vector the electrical current. Hereby $\sigma_1, \sigma_2, \sigma_3$ are to be identified with the components i_x, i_y, i_z of the electrical current, and

$\sigma_4 = i\sigma$ indicates, with σ being the charge density, the density of the actual electricity of the Maxwell-Hertz theory. This vector satisfies the continuity equation $\text{Div}(\sigma) = 0$. In addition, I introduce another thing which for now, if you don't mind, I shall call a *Traktor*. The latter shall be a formation consisting of six components which, in a four dimensional space, become attached to the concept of the vector, namely,

$$p_{23}, \quad p_{31}, \quad p_{12}, \quad p_{14}, \quad p_{24}, \quad p_{34}.$$

A Traktor carries, like wise, significance which does not depend on the choice of a four dimensional coordinate system. That is to say that new orthogonal coordinates for x_1, x_2, x_3, x_4, namely, y_1, y_2, y_3, y_4 are inserted in such a way that the latter quadruple of variables remains congruent with the former quadruple, that is, through this linear substitution each set of variables is transformable into the other and, moreover, at the same time the values of p_{jk} are to be substituted by this transformation in such a way that they correspond with the coefficients of the bilinear expression

$$p_{23}(x_2 y_3 - x_3 y_2) + p_{31}(x_3 y_1 - x_1 y_3) + \ldots + p_{34}(x_3 y_4 - x_4 y_3).$$

I shall now describe the fundamental equations of the electrodynamics of moving mediums: The following points have to be taken into consideration: For every spacetime point x, y, z, t there is a corresponding vector potential $(\psi) = \psi_1, \psi_2, \psi_3, \psi_4$ of the electromagnetic field, further a velocity vector $(w) = w_1, w_2, w_3, w_4$ of matter conforms to this, as does the vector of the electric current given by $(\sigma) = \sigma_1, \sigma_2, \sigma_3, \sigma_4$, and finally a Traktor, which I will call a Polarisationstraktor, $(p) = p_{23}, \ldots p_{34}$ comes into play.

The vector (ψ), although it is not grounded in a four dimensional space, must satisfy the fundamental equation $\text{Div}(\psi) = 0$. From this vector arises, as shown above, a Traktor $\psi_{23}, \ldots \psi_{34}$, whereby now the first three components indicate the magnetic induction and the following three the electrical field strength. In addition we already have the differential equation,

$$\frac{\partial p_{1j}}{\partial x_1} + \frac{\partial p_{2j}}{\partial x_2} + \frac{\partial p_{3j}}{\partial x_3} + \frac{\partial p_{4j}}{\partial x_4} = \sigma_j + \square \psi_j \ldots \qquad [p_{kj} = -p_{jk}].$$

Finally the following facts are valid:

If one transforms the components in such a way that a determined position x, y, z, t stays at rest, then p_{14}, p_{24}, p_{34}, multiplied by i, becomes itself the dielectric polarization in proportion to the vector $\psi_{14}, \psi_{24}, \psi_{34}$, whereby the proportionality factor decreases the dielectric constant of the medium by 1; furthermore, with regard to objects that cannot be magnetized, the vector p_{23}, p_{31}, p_{12} is in that case zero; finally, the vector $\sigma_1, \sigma_2, \sigma_3$ is also proportional to the vector $\psi_{14}, \psi_{24}, \psi_{34}$, and the proportionality factor is the electrical capacity of conductivity. Evidently, under these terms or stipulations the invariance of the *Lorentzian* group rests ascertained.

3. Now we will turn to mechanics. According to what I have already said about the relationship between the relativity principle and the law of inertia, it should, for a start, be clearly understood that the prevailing fundamental laws of mechanics count for nothing more than an approximation of reality if, with regard to mechanics, the postulate of relativity is to be validated. But such has to be again the case because otherwise the possibility that the earth moves relative to the aether would have to be reinstated. *Planck* clearly points to the necessity of dismissal of the law of inertia under the following consideration. If one takes thermal radiation into account, then it becomes impossible to separate the energy of a uniformly moving body from any other energy of that body. In general, one cannot define the kinetic energy of a body in a determined manner, that is, there always remains a gap in the laws of mechanics which can only be filled by applying the relativity principle. If we think of an empty space or cavity surrounded by a very expansive and perfect conductor, then, on account of its huge volume, the influence of the matter of the walls of this cavity can further be considered negligible. Accordingly, if inside this cavity the thermal radiation is designated by some temperature T, and if, thereby, a body moves with a constant velocity w in any direction, then the volume of this cavity can finally be designated by V. Hence, the theory that accounts for the energy contained in radiation can be formulated as:

$$ E = \frac{1 + \frac{w^2}{8}}{(1 - w^2)^3} T^4 V, $$

however, a suitable temperature unit must be chosen for this application. Now, provided that one keeps in mind that w^4 stands against 1, one cannot determine by this mathematical expression a term for a state of rest and a proportional term for w^2, that is, one cannot separate the kinetic energy of uniform motion from the inherent energy of a body. Moreover, such a separation, in the context of the energy of thermal radiation, is naturally and generally in all cases impossible. For example, *Planck* calculates that, for a contained space filled with a stationary mono-atomic gas at the melting point temperature of platinum and under a pressure of 0,001 mm, the heat that is added under constant pressure accounts for a fourth part of the increase in radiation energy, that is, molecular motion accounts but for three quarters of the radiation energy thus generated. Apparently then, the first quarter of this energy cannot be ignored. In his recently published article in the *Berliner Akademie* (18. June of this year), *Planck* tries to establish a dynamic formulation on the basis of the relativity principle. As a further goal plausible mathematical attempts should be worked out in such a way that experimental tests can be performed and interpreted under the new as well as past theories. *Planck*'s concern is the dynamic behaviour of a single point when it is subjected to a change of temperature. This shows that if, at times, one turns away from cosmic events, then thermodynamic influences make it possible to detect important phenomena in the first place. *Planck*'s thoughts encompass a body which on the whole only experiences translational motions, that is, a body whose motions can be determined by three velocity components, namely, w_x, w_y, w_z, and whose state depends in addition on its volume V and its temperature T. Hence, what *Planck* is searching for are the laws of the dynamics of such a system. The above mentioned body is, however, at the same time in a state of rest with regard to a certain coordinate system; the question is, at what temperature is this body in a state of rest? Changes in volume are given by this substitution of the coordinate system. Now *Planck* plausibly explains that the entropy with respect to both coordinate systems must remain the same; accordingly then, with regard to a thermal radiation filled cavity, we can in this case apply the known laws, with constant pressure taken into consideration, so that T in $T/\sqrt{1 - w^2}$ undergoes a change with respect to the new coordinate system. *Planck* accepts that this relationship holds generally true on account that pressure and temperature in this case present equilibrium parameters. Still missing are the laws under which this system is subjected by given influential outside forces in terms of pressure and temperature, and therefore *Planck* accepts *Helmholtz'* extension of the principle of least action which can be expanded to also hold true for thermodynamic relationships. The change between two states, at a constant energy level, should therefore be given by the minimum of the integral

$$W = \int H \mathrm{d}t,$$

whereby H, the so called kinetic potential, depends on the three velocity components, yet this holds only in connection with w, as well as with the volume V and the temperature T.

The result, which *Planck* arrives at via a detour by dragging thermodynamic relations into place, could simply have been foreseen through the equation

$$H \, dt = K \, \sqrt{dt^2 - dx^2 - dy^2 - dz^2},$$

whereby the second factor of the *Lorentzian* group remains invariant, which means, that the first factor stays invariant as well. The noteworthy conclusion that now ties up with this is that the transferral motion of the body, which changes with time and thus presents the external force, can be established by the following equations:

$$G = \frac{\partial H}{\partial w} = \frac{w(E + pV)}{c^2}$$

whereby

$$E = w\frac{\partial H}{\partial w} + T\frac{\partial H}{\partial T} - H$$

gives us the total energy of the body; I wrote the above mathematical expression, which does not specify any particular unit of time, in order to add the following remark. If now the velocity changes only in accord with a transversal vector and the affiliated change in size of the motion is M times the acceleration, then M denotes an increase in size which can be called the mass of the body. Therefore:

$$M = \frac{G}{w} = \frac{E + pV}{c^2}.$$

Accordingly: Considering that mass denotes the size of the body, what matters here in particular is that the quantity of mass changes with any addition of heat. That is, by an addition of heat under constant pressure the increase in mass amounts exactly to the added heat divided by the square of the speed of light. For example, if 18 g (1 $\frac{1}{2}$ mol) of oxy-hydrogen gas under atmospheric pressure and at room temperature condenses to form liquid water, then a decrease of mass of 3.2 10^{-6} mg should occur, a diminution of matter so small, however, that we cannot observe it. It would naturally stand to reason, nevertheless, that a further development of this theory will consequently allow us to observe this phenomenon.

4. Finally I want to say a few words about gravitation. The important question remains, how the law of gravitation pertains to the realm of the principle of relativity. In his article submitted to the *Rendiconti del Circolo Matematico di Palermo* in 1906, Poincaré addressed this question. Laplace believed that he had proven that the propagation of gravitation must happen momentarily or at a much higher speed than the speed of light. Yet, if one were to ground gravitation in electromagnetism, then Laplace's proof would

surely emerge as incomplete. In fact, Poincaré as well thinks that gravitation propagates in no other way than at the speed of light. Apparently then, the purely mathematical task consists in finding a law which corresponds to the relativity principle and which, at the same time, must be applicable to Newton's laws, that is, if one can neglect the squares of the speed of stars as well as the multiplied result of outward acceleration (i.e., proportional to the square of the speed of light). Poincaré points out such a law by taking into consideration the invariants of the Lorentzian group, however, this law is but one amongst many other possible laws, and the relevant investigations are in no way of a definitive character. Perhaps I will write in greater detail about this at another time. I hope my above account has made one thing clear, namely, that the current investigations should claim, or even demand, the highest degree of interest of mathematicians.

(Received, June 15, 1915)

Footnote on the first page of the original German publication of *The Relativity Principle* in 1915 (H. Minkowski, Das Relativitätsprinzip, *Annalen der Physik* **47** (1915) S. 927-938):

Exactly 10 years ago the *Annalen der Physik* published Einstein's work about the electrodynamics of moving bodies. With respect to the important role of the relativity principle, based upon this work and played out in the *Annalen*, Mr. Sommerfeld, in agreement with the editorial department of the *Annalen* and from Minkowski's posthumous papers, as the most successful interpretation of the relativity principle, publicly presented the following account. This lecture was given at the meeting of the *Göttinger Mathematischen Gesellschaft* on November 5, 1907, that is, almost one year before Minkowski's lecture on space and time that took place in Cologne.

The Fundamental Equations for Electromagnetic Processes in Moving Bodies

Introduction

At present differences of opinion on the basic equations of electrodynamics for moving bodies are still prevailing. The approach of *Hertz*[1] (1890) has to be abandoned because it has been found that it contradicts various experimental results.

In 1895 *H. A. Lorentz*[2] published his theory of optical and electrical phenomena in moving bodies, which was based on an atomistic understanding of electricity, and whose many successes seem to have justified the bold hypotheses. Lorentz' theory[3] assumes some initial equations, which should be valid at every point of "aether"; then by forming the average values over "physically infinitely small" regions that already contain many "electrons," the equations for electromagnetic processes in moving bodies can be obtained.

In particular, Lorentz' theory gives an account of the non-existence of a relative motion of the Earth with respect to the luminiferous aether; it brings this fact in connection with a covariance of those initial equations with certain simultaneous transformations of the space and time parameters, which have received from *H. Poincaré*[4] the name *Lorentz transformations*.

---

[1] Über die Gründgleichungen der Elektrodynamik für bewegte Körper. Wiedemanns Annalen, Bd. 41, S. 369, 1890 (auch in: Gesammelte Werke, Bd. I, S. 256, Leipzig 1892).

[2] Versuch einer Theorie der elektrischen und optischen Erscheinungen in be- wegten Körpern, Leiden 1895.

[3] Vgl. Enzyklopädie der mathematischen Wissenschaften, V 2, Art. 14. Weiterbildung der Maxwellschen Theorie. Elektronentheorie.

[4] Rendiconti del Circolo Matematico di Palermo, T. XXI (1906), p. 129.

For those initial equations, the covariance under the Lorentz transformations is a purely mathematical fact, which I will call the *theorem of relativity*; this theorem is essentially based on the form of the differential equation for the propagation of waves with the velocity of light.

It is now possible without any hypothesis about the connection between electricity and matter, to expect that this mathematically evident theorem will have its consequences extended so far that they may hold even for those laws of ponderable media which are yet unknown, and which may possess this covariance under Lorentz transformations. This expresses therefore more a confidence than already an existing understanding, and this confidence I will call the *postulate of relativity*. This situation is approximately such, as if one postulates the conservation of energy in cases where the common forms of energy are still not recognized.

If afterwards the expected covariance is manifested as a specific relation between observable quantities for moving bodies, this particular relation may then be called the *principle of relativity*.

These distinctions seem to me useful and can characterize the current state of the electrodynamics of moving bodies.

H. A. Lorentz found the relativity theorem and established the relativity postulate as a hypothesis that electrons and matter experience contractions as a result of their motion according to a certain law.

A. Einstein[5] expressed it most sharply so far that this *postulate* is not an artificial hypothesis but is rather a new concept of time imposed upon us by the phenomena.

However, the *Principle* of Relativity, in the sense indicated by me, has not yet been formulated for the electrodynamics of moving bodies. *As I formulate this principle in the present paper I obtain the fundamental equations for moving bodies in a form entirely and unambiguously determined by this principle. We will see that none of the previously accepted forms for these equations exactly comply with this principle.*

One would particularly expect that the adopted by Lorentz fundamental equations for moving bodies correspond to the relativity postulate. It turns out however, that this is not the case for the general equations derived by Lorentz for an arbitrary, also magnetized, body but this is the case *approximately* (neglecting the squares of the speeds of matter against the square of the speed of light) for those equations which Lorentz deduced afterwards for a non-magnetic body; but this later adjustment to the relativity postulate, in view of the fact that the condition of non-magnetization has been for-

---

[5]Annalen der Physik, Bd. 17, S. 891, 1905.

mulated in a way not corresponding to the relativity postulate, is achieved by an accidental compensation of two violations to the relativity postulate. However, this statement means no objection against the molecular hypothesis of Lorentz, but it is only clear that the adoption of the contraction of the electrons in motion in the Lorentz' theory would have to be introduced at an earlier point than this was done by Lorentz.

In an appendix I discuss the status of classical mechanics with respect to the relativity postulate. An easy change of mechanics in accordance with the requirements of the postulate of relativity would give for the observed phenomena hardly noticeable differences, but would lead to very surprising results: *Starting with the relativity postulate, there are sufficient means for deducing henceforth the complete set of laws of mechanics from the law of conservation of energy alone* (and statements about the forms of energy).

§1.

Notations

Let a reference system of rectangular coordinates x, y, z, t be given in space and time. The unit of time is selected in such a relation to the unit of length that the velocity of light in empty space is 1.

Although I would prefer not to change the notations used by *Lorentz* it seems important to me to use from the very beginning a different selection of symbols and to establish their relationships. I will call the vector

> *electric force by* $\mathfrak{E}$, *the magnetic induction by* $\mathfrak{M}$, *the electric induction by* $\mathfrak{e}$ *and the magnetic force by* $\mathfrak{m}$,

so that $\mathfrak{E}$, $\mathfrak{M}$, $\mathfrak{e}$, $\mathfrak{m}$ are used instead of $\mathfrak{E}$, $\mathfrak{B}$, $\mathfrak{D}$, $\mathfrak{H}$ used by Lorentz.

I will also use the complex variables in a manner that has not been previously common practice in physical studies, i.e., instead of operating with t, I will operate with it, where i is the imaginary unit $\sqrt{-1}$. On the other hand, very important circumstances become evident as I use a notation with indices, namely, often instead of

$$x, \ y, \ z, \ it \qquad\qquad x_1, \ x_2, \ x_3, \ x_4$$

is used which will then establish a general use of the indices 1, 2, 3, 4. The advantage of this method will be, as I explicitly emphasize here, that it only provides a more evident expression of *purely real relations*; we can however at any moment go to real equations if it is assumed that symbols with *one* index 4 represent *purely imaginary* values, while those which *have no* index 4, or have it *twice* represent *real* values.

An individual system of values of x, y, z, t, i. e. of x_1, x_2, x_3, x_4 will be called a *spacetime point*.

Further let $\mathfrak{w}$ denote the *velocity vector of matter*, ε the *dielectric constant*, μ, the *magnetic permeability*, σ the *conductivity of matter*, which all are functions of x, y, z, t (or x_1, x_2, x_3, x_4), while ρ denotes the *density of electricity in space*, and $\mathfrak{s}$ the vector of "electric current" which we will introduce in §7 and §8.

First Part.
Consideration of aether as a limiting case

§2.
The basic equations for ether

By using the electron theory, Lorentz in his above mentioned essay traces the laws of electrodynamics of ponderable bodies to still simpler laws. Let us now adhere to these simpler laws, whereby we require that for the limiting case $\varepsilon = 1$, $\mu = 1$, $\sigma = 0$, they should constitute the laws for ponderable bodies. In this ideal limiting case $\varepsilon = 1$, $\mu = 1$, $\sigma = 0$, $\mathfrak{E} = \mathfrak{e}$, and $\mathfrak{M} = \mathfrak{m}$. At every space time point x, y, z, t we shall have the equations

$$\operatorname{curl} \mathfrak{m} - \frac{\partial \mathfrak{e}}{\partial t} = \varrho \mathfrak{w}, \tag{I}$$

$$\operatorname{div} \mathfrak{e} = \varrho, \tag{II}$$

$$\operatorname{curl} \mathfrak{e} + \frac{\partial \mathfrak{m}}{\partial t} = 0, \tag{III}$$

$$\operatorname{div} \mathfrak{m} = 0. \tag{IV}$$

I will now write x_1, x_2, x_3, x_4 for x, y, z, it $(i = \sqrt{-1})$ and

$$\varrho_1, \ \varrho_2, \ \varrho_3, \ \varrho_4$$

for

$$\varrho\mathfrak{w}_x, \quad \varrho\mathfrak{w}_y, \quad \varrho\mathfrak{w}_z, \quad i\varrho$$

i.e. the components of the convection current $\varrho\mathfrak{w}$ and the multiplied by i electric density. Further I will write

$$f_{23}, \quad f_{31}, \quad f_{12}, \quad f_{14}, \quad f_{24}, \quad f_{34}$$

for

$$\mathfrak{m}_x, \quad \mathfrak{m}_y, \quad \mathfrak{m}_z, \quad -i\mathfrak{e}_x, -i\mathfrak{e}_y, -i\mathfrak{e}_z$$

i.e., the components of $\mathfrak{m}$ and $-i\mathfrak{e}$ along the three axes; finally for any two indices h, k of the numbers 1, 2, 3, 4

$$f_{kh} = -f_{hk},$$

therefore

$$f_{32} = -f_{23}, \; f_{13} = -f_{31}, \; f_{21} = -f_{12},$$

$$f_{41} = -f_{14}, \; f_{42} = -f_{24}, \; f_{43} = -f_{34}.$$

Then the three equations in (I), and the equation (II) multiplied by i become:

$$\frac{\partial f_{12}}{\partial x_2} + \frac{\partial f_{13}}{\partial x_3} + \frac{\partial f_{14}}{\partial x_4} = \varrho_1,$$

$$\frac{\partial f_{21}}{\partial x_1} \qquad + \frac{\partial f_{23}}{\partial x_3} + \frac{\partial f_{24}}{\partial x_4} = \varrho_2,$$

$$\frac{\partial f_{31}}{\partial x_1} + \frac{\partial f_{32}}{\partial x_2} \qquad + \frac{\partial f_{34}}{\partial x_4} = \varrho_3, \qquad \text{(A)}$$

$$\frac{\partial f_{41}}{\partial x_1} + \frac{\partial f_{42}}{\partial x_2} + \frac{\partial f_{43}}{\partial x_3} \qquad = \varrho_4.$$

On the other hand, the three equations in (III) multiplied by $-i$, and the equation (IV) multiplied by -1, become

$$\frac{\partial f_{34}}{\partial x_2} + \frac{\partial f_{42}}{\partial x_3} + \frac{\partial f_{23}}{\partial x_4} = 0,$$

$$\frac{\partial f_{43}}{\partial x_1} \qquad + \frac{\partial f_{14}}{\partial x_3} + \frac{\partial f_{31}}{\partial x_4} = 0,$$

$$\frac{\partial f_{24}}{\partial x_1} + \frac{\partial f_{41}}{\partial x_2} \qquad + \frac{\partial f_{12}}{\partial x_4} = 0, \qquad \text{(B)}$$

$$\frac{\partial f_{32}}{\partial x_1} + \frac{\partial f_{13}}{\partial x_2} + \frac{\partial f_{21}}{\partial x_3} \qquad = 0.$$

One notices immediately in this way of writing the *perfect symmetry* of the 1st as well as the 2nd system of equations *with respect to permutation of the indices* $1, 2, 3, 4$.

§3.

The Theorem of Relativity of Lorentz

It is well-known that writing the equations (I)–(IV) in vectorial notation reveals an invariance (or rather a covariance) of the system of equations (A) as well as of (B), when the coordinate system is rotated around the origin. For example, if we take a rotation of the axes about the $z-$axis at an angle φ, keeping $\mathfrak{e}, \mathfrak{m}, \mathfrak{w}$ fixed in space, and instead of x_1, x_2, x_3, x_4 introduce new variables x_1', x_2', x_3', x_4' by

$$x_1' = x_1 \cos\varphi + x_2 \sin\varphi, \quad x_2' = -x_1 \sin\varphi + x_2 \cos\varphi,$$

$$x_3' = x_3, \quad x_4' = x_4,$$

and introduce new quantities ϱ_1', ϱ_2', ϱ_3', ϱ_4' by

$$\varrho_1' = \varrho_1 \cos\varphi + \varrho_2 \sin\varphi, \quad \varrho_2' = -\varrho_1 \sin\varphi + \varrho_2 \cos\varphi,$$

$$\varrho_3' = \varrho_3, \quad \varrho_4' = \varrho_4,$$

and also new quantities $f_{12}', \cdots, f_{34}'$ by

$$f_{23}' = f_{23} \cos\varphi + f_{31} \sin\varphi, \quad f_{31}' = -f_{23} \sin\varphi + f_{31} \cos\varphi, \quad f_{12}' = f_{12},$$

$$f_{14}' = f_{14} \cos\varphi + f_{24} \sin\varphi, \quad f_{24}' = -f_{14} \sin\varphi + f_{24} \cos\varphi, \quad f_{34}' = f_{34},$$

$$f_{kh}' = -f_{hk}' \qquad (h, k = 1, 2, 3, 4),$$

then out of the equations (A) would follow a corresponding system of primed equations (A′) composed of the newly introduced primed quantities.

Now the theorem of relativity, which was discovered by Lorentz, follows immediately without any calculation from the symmetry of the equations (A) and (B) with respect to the indices $1, 2, 3, 4$.

I will understand by $i\psi$ a purely imaginary quantity and will consider the substitution

$$x_1' = x_1, \ x_2' = x_2, \ x_3' = x_3 \cos i\psi + x_4 \sin i\psi, \ x_4' = -x_3 \sin i\psi + x_4 \cos i\psi.$$

$$(1)$$

By means of

$$-i \, \text{tg} \, i\psi = \frac{e^\psi - e^{-\psi}}{e^\psi + e^{-\psi}} = q, \quad \psi = \frac{1}{2} \log \text{nat} \, \frac{1+q}{1-q} \tag{2}$$

we have

$$\cos i\psi = \frac{1}{\sqrt{1-q^2}}, \quad \sin i\psi = \frac{iq}{\sqrt{1-q^2}},$$

where $-1 < q < 1$ and $\sqrt{1-q^2}$ is taken with positive sign. By writing

$$x_1' = x', \quad x_2' = y', \quad x_3' = z', \quad x_4' = it', \tag{3}$$

the substitution (1) takes the form

$$x' = x, \quad y' = y, \quad z' = \frac{z - qt}{\sqrt{1-q^2}}, \quad t' = \frac{-qz + t}{\sqrt{1-q^2}} \tag{4}$$

with *only real coefficients*.

If now in the above equation describing a rotation about the $z-$ axis, we replace $1, 2, 3, 4$ throughout by $3, 4, 1, 2$, and φ by $i\psi$, and at the same time introduce new quantities ϱ_1', ϱ_2', ϱ_3', ϱ_4' through the substitution

$$\varrho_1' = \varrho_1, \quad \varrho_2' = \varrho_2, \quad \varrho_3' = \varrho_3 \cos i\psi + \varrho_4 \sin i\psi,$$

$$\varrho_4' = -\varrho_3 \sin i\psi + \varrho_4 \cos i\psi,$$

and also new quantities $f_{12}', \cdots, f_{34}'$ introduced by

$$f_{41}' = f_{41} \cos i\psi + f_{13} \sin i\psi, \quad f_{13}' = -f_{41} \sin i\psi + f_{13} \cos i\psi, \quad f_{34}' = f_{34},$$

$$f_{32}' = f_{32} \cos i\psi + f_{42} \sin i\psi, \quad f_{42}' = -f_{32} \sin i\psi + f_{42} \cos i\psi, \quad f_{12}' = f_{12},$$

$$f_{kh}' = -f_{hk}' \qquad (h, k = 1, 2, 3, 4),$$

then the systems of equations in (A) and (B) are transformed into equations (A'), and (B'), and we see that the new equations can be obtained by simply priming the old equation.

All these equations can be now rewritten in a purely real form and the final result can be formulated in the following way:

If the real transformations (4) are used, and x', y', z', t' are regarded as a reference system for space and time, and we introduce[6]

$$\varrho' = \varrho \left(\frac{-q\mathfrak{w}_z + 1}{\sqrt{1-q^2}} \right), \quad \varrho' \mathfrak{w}_z' = \varrho \left(\frac{\mathfrak{w}_z - q}{\sqrt{1-q^2}} \right), \tag{5}$$

---

[6]The equations (5) are here in a different order, whereas the equations (6) and (7) are in the same sequence as the above equations, from which they follow.

$$\varrho'\mathfrak{w}'_{x'} = \varrho\mathfrak{w}_x, \quad \varrho'\mathfrak{w}'_{y'} = \varrho\mathfrak{w}_y,$$

further

$$\mathfrak{e}'_{x'} = \frac{\mathfrak{e}_x - q\mathfrak{m}_y}{\sqrt{1 - q^2}}, \quad \mathfrak{m}'_{y'} = \frac{-q\mathfrak{e}_x + \mathfrak{m}_y}{\sqrt{1 - q^2}}, \quad \mathfrak{e}'_{z'} = \mathfrak{e}_z \tag{6}$$

and

$$\mathfrak{m}'_{x'} = \frac{\mathfrak{m}_x + q\mathfrak{e}_y}{\sqrt{1 - q^2}}, \quad \mathfrak{e}'_{y'} = \frac{q\mathfrak{m}_x + \mathfrak{e}_y}{\sqrt{1 - q^2}}, \quad \mathfrak{m}'_{z'} = \mathfrak{m}_z, \tag{7}$$

then for the vectors $\mathfrak{w}', \mathfrak{e}', \mathfrak{m}'$ with components $\mathfrak{w}'_{x'}, \mathfrak{w}'_{y'}, \mathfrak{w}'_{z'}; \mathfrak{e}'_{x'}, \mathfrak{e}'_{y'}, \mathfrak{e}'_{z'};$ $\mathfrak{m}'_{x'}, \mathfrak{m}'_{y'}, \mathfrak{m}'_{z'}$ in the new coordinate system x', y', z', and also for the quantity ϱ' the equations (I')–(IV'), which are analogous to (I)–(IV), are exactly satisfied, whereas the system (I), (II) goes into (I'), (II') and the system (III), (IV) into (III'), (IV').

We remark that here $\mathfrak{e}_x - q\mathfrak{m}_y, \mathfrak{e}_y + q\mathfrak{m}_x, \mathfrak{e}_z$, are the components of the vector $\mathfrak{e} + [\mathfrak{b}\,\mathfrak{m}]$, where $\mathfrak{b}$ is a vector in the direction of the positive $z-$axis, and is of absolute value $|\mathfrak{b}| = q$, and $[\mathfrak{b}\,\mathfrak{m}]$ is the vector product of $\mathfrak{b}$ and $\mathfrak{m}$. Analogously $\mathfrak{m}_x + q\mathfrak{e}_y, \mathfrak{m}_y - q\mathfrak{e}_x, \mathfrak{m}_z$ are the components of the vector $\mathfrak{m} - [\mathfrak{b}\,\mathfrak{e}]$.

The equations (6) and (7), as they are in pairs *with each other*, can be combined by another use of imaginary quantities in

$$\begin{aligned}
\mathfrak{e}'_{x'} + i\mathfrak{m}'_{x'} &= (\mathfrak{e}_x + i\mathfrak{m}_x)\cos i\psi + (\mathfrak{e}_y + i\mathfrak{m}_y)\sin i\psi, \\
\mathfrak{e}'_{y'} + i\mathfrak{m}'_{y'} &= -(\mathfrak{e}_x + i\mathfrak{m}_x)\sin i\psi + (\mathfrak{e}_y + i\mathfrak{m}_y)\cos i\psi, \\
\mathfrak{e}'_{z'} + i\mathfrak{m}'_{z'} &= \mathfrak{e}_z + i\mathfrak{m}_z.
\end{aligned}$$

and we notice that if φ is any real angle, from the these relationships the following combinations can be formed:

$$\begin{aligned}
(\mathfrak{e}'_{x'} + i\mathfrak{m}'_{x'})\cos\varphi &+ (\mathfrak{e}'_{y'} + i\mathfrak{m}'_{y'})\sin\varphi \\
&= (\mathfrak{e}_x + i\mathfrak{m}_x)\cos(\varphi + i\psi) + (\mathfrak{e}_y + i\mathfrak{m}_y)\sin(\varphi + i\psi), \quad (8)
\end{aligned}$$

$$\begin{aligned}
-(\mathfrak{e}'_{x'} + i\mathfrak{m}'_{x'})\sin\varphi &+ (\mathfrak{e}'_{y'} + i\mathfrak{m}'_{y'})\cos\varphi \\
&= -(\mathfrak{e}_x + i\mathfrak{m}_x)\sin(\varphi + i\psi) + (\mathfrak{e}_y + i\mathfrak{m}_y)\cos(\varphi + i\psi). \quad (9)
\end{aligned}$$

§4.

Special Lorentz transformations

The role which is played by the $z-$axis in the transformation (4) can easily be transferred to any other direction. Each of the systems of coordinate axes x, y, z and x', z', y' is subjected to the same rotation with respect to itself. We come now to a more general proposition.

Let $\mathfrak{b}$ be a vector in any direction with components $\mathfrak{b}_x, \mathfrak{b}_y, \mathfrak{b}_z$, and with such a non-zero absolute value $|\mathfrak{b}| = q$ which *is smaller than* 1. By $\overline{\mathfrak{b}}$ we shall denote any vector which is perpendicular to $\mathfrak{b}$, and further denote the components of a vector $\mathfrak{r}$ in the direction of $\mathfrak{b}$ and $\overline{\mathfrak{b}}$ as $\mathfrak{r}_{\mathfrak{b}}$ and $\mathfrak{r}_{\overline{\mathfrak{b}}}$, respectively.

Instead of x, y, z, t, new variables x', y', z', t' will be introduced in the following way. If for the sake of shortness, $\mathfrak{r}$ is written for the vector with the components x, y, z in the first system of reference, $\mathfrak{r}'$ for the same vector with the components x', y', z' in the second system of reference, then *for the direction of $\mathfrak{b}$* we have

$$\mathfrak{r}'_{\mathfrak{b}} = \frac{\mathfrak{r}_{\mathfrak{b}} - qt}{\sqrt{1 - q^2}}, \tag{10}$$

and for each $\overline{\mathfrak{b}}$ in a direction perpendicular to $\mathfrak{b}$

$$\mathfrak{r}'_{\overline{\mathfrak{b}}} = \mathfrak{r}_{\overline{\mathfrak{b}}}, \tag{11}$$

and also

$$t' = \frac{-q\mathfrak{r}_{\mathfrak{b}} + t}{\sqrt{1 - q^2}}. \tag{12}$$

The notations $\mathfrak{r}'_{\mathfrak{b}}$ and $\mathfrak{r}'_{\overline{\mathfrak{b}}}$ should be understood in the sense that the direction of $\mathfrak{b}$ and every direction $\overline{\mathfrak{b}}$ perpendicular to $\mathfrak{b}$ in the system x, y, z a direction is always associated with the same direction cosines in the system x', y', z'.

A transformation which is represented by (10), (11), (12) with the condition $0 < q < 1$ I will call a *special Lorentz transformation*, and $\mathfrak{b}$ will be called the *vector*, the direction of $\mathfrak{b}$ – the *axis*, and the magnitude of $\mathfrak{b}$ – the *moment* of this special Lorentz transformation.

If further ϱ' and the vectors $\mathfrak{w}', \mathfrak{e}', \mathfrak{m}',$ in the system x', y', z' are so defined that,

$$\varrho' = \frac{\varrho(-q\mathfrak{w}_{\mathfrak{b}} + 1)}{\sqrt{1 - q^2}}, \tag{13}$$

$$\varrho'\mathfrak{w}'_{\mathfrak{b}} = \frac{\varrho\mathfrak{w}_{\mathfrak{b}} - \varrho q}{\sqrt{1 - q^2}}, \quad \varrho'\mathfrak{w}'_{\overline{\mathfrak{b}}} = \varrho\mathfrak{w}_{\overline{\mathfrak{b}}} \tag{14}$$

and further[7]

$$(\mathfrak{e}' + i\mathfrak{m}')_{\overline{\mathfrak{b}}} = \frac{(\mathfrak{e} + i\mathfrak{m} - i[\mathfrak{w}, \mathfrak{e} + i\mathfrak{m}])_{\overline{\mathfrak{b}}}}{\sqrt{1 - q^2}},$$

$$(\mathfrak{e}' + i\mathfrak{m}')_{\mathfrak{b}} = (\mathfrak{e} + i\mathfrak{m} - i[\mathfrak{w}, \mathfrak{e} + i\mathfrak{m}])_{\mathfrak{b}}, \qquad (15)$$

then it follows that the systems of equations (I), (II), and (III), (IV) every time are transformed into the exactly corresponding systems with primes.

The solution of the equations (10), (11), (12) leads to

$$\mathfrak{r}_{\mathfrak{b}} = \frac{\mathfrak{r}'_{\mathfrak{b}} + qt'}{\sqrt{1 - q^2}}, \quad \mathfrak{r}_{\overline{\mathfrak{b}}} = \mathfrak{r}'_{\overline{\mathfrak{b}}}, \quad t = \frac{q\mathfrak{r}'_{\mathfrak{b}} + t'}{\sqrt{1 - q^2}}. \qquad (16)$$

We shall now make a very important observation about the vectors $\mathfrak{w}$ and $\mathfrak{w}'$. We can again introduce the indices $1, 2, 3, 4$, so that we write x'_1, x'_2, x'_3, x'_4 instead of x', y', z', it' and ϱ'_1, ϱ'_2, ϱ'_3, ϱ'_4 instead of $\varrho'\mathfrak{w}'_{x'}$, $\varrho'\mathfrak{w}'_{y'}$, $\varrho'\mathfrak{w}'_{z'}$, $i\varrho'$. Like the rotation about the z−axis, the transformation (4), and the more general transformations (10), (11), (12), are also *linear transformations* with the determinant $+1$, *with the help of which*

$$x_1^2 + x_2^2 + x_3^2 + x_4^2, \quad \text{i.e.} \quad x^2 + y^2 + z^2 - t^2 \qquad (17)$$

is transformed into

$$x_1'^2 + x_2'^2 + x_3'^2 + x_4'^2, \quad \text{i.e.} \quad x'^2 + y'^2 + z'^2 - t'^2.$$

On the basis of the equations (13), (14)

$$-(\varrho_1^2 + \varrho_2^2 + \varrho_3^2 + \varrho_4^2) = \varrho^2(1 - \mathfrak{w}_x^2 - \mathfrak{w}_y^2 - \mathfrak{w}_z^2) = \varrho^2(1 - \mathfrak{w}^2)$$

is transformed into $\varrho'^2(1 - \mathfrak{w}'^2)$ or in other words the square root of it

$$\varrho\sqrt{1 - \mathfrak{w}^2}, \qquad (18)$$

which is positive, is an *invariant* under the Lorentz transformations.

If we divide ϱ_1, ϱ_2, ϱ_3, ϱ_4 by this magnitude, we obtain the 4 values

$$w_1 = \frac{\mathfrak{w}_x}{\sqrt{1 - \mathfrak{w}^2}}, \quad w_2 = \frac{\mathfrak{w}_y}{\sqrt{1 - \mathfrak{w}^2}}, \quad w_3 = \frac{\mathfrak{w}_z}{\sqrt{1 - \mathfrak{w}^2}}, \quad w_4 = \frac{i}{\sqrt{1 - \mathfrak{w}^2}},$$

between which the relationship

$$w_1^2 + w_2^2 + w_3^2 + w_4^2 = -1 \qquad (19)$$

---

[7]The parentheses only combine the terms with the same index, whereas $[\mathfrak{w}, \mathfrak{e} + i\mathfrak{m}]$ is the vector product of $\mathfrak{w}$ and $\mathfrak{e} + i\mathfrak{m}$.

exists. Obviously these 4 values are uniquely determined by the vector $\mathfrak{w}$ and vice versa – the quantities w_1, w_2, w_3, w_4, where w_1, w_2, w_3 are real and $-iw_4$ is real and positive, through the condition (19) uniquely determine the vector $\mathfrak{w}$ with absolute value < 1.

The meaning of w_1, w_2, w_3, w_4 here is that they are the ratios of dx_1, dx_2, dx_3, dx_4 to

$$\sqrt{-(dx_1^2 + dx_2^2 + dx_3^2 + dx_4^2)} = dt\sqrt{1 - \mathfrak{w}^2}. \tag{20}$$

where the differentials are the displacements of matter occupying the space-time point x_1, x_2, x_3, x_4 to an adjacent spacetime point. Now the equations (10), (11), (12) immediately apply to the matching differentials dx, dy, dz, dt and dx', dy', dz', dt' and will therefore lead to

$$-(dx_1^2 + dx_2^2 + dx_3^2 + dx_4^2) = -(dx_1'^2 + dx_2'^2 + dx_3'^2 + dx_4'^2).$$

After performing the Lorentz transformation the velocity of matter in the new system of reference for the same spacetime point x', y', z', t' is the vector $\mathfrak{w}'$ with the ratios $\frac{dx'}{dt'}$, $\frac{dy'}{dt'}$, $\frac{dz'}{dt'}$ interpreted as components.

Now it is evident that the system of variables

$$x_1 = w_1, \quad x_2 = w_2, \quad x_3 = w_3, \quad x_4 = w_4$$

by virtue of the Lorentz transformation (10), (11), (12) transforms in the new system of variables

$$x_1' = w_1', \quad x_2' = w_2', \quad x_3' = w_3', \quad x_4' = w_4',$$

which has exactly the same meaning for the velocity $\mathfrak{w}'$ *after* the transformation as the former system of variables for the velocity $\mathfrak{w}$ *before* the transformation.

If in particular the vector $\mathfrak{b}$ of the special Lorentz transformation is equal to the velocity vector $\mathfrak{w}$ of matter at the spacetime point x', y', z', t' then it follows from (10), (11), (12) that

$$w_1' = 0, \quad w_2' = 0, \quad w_3' = 0, \quad w_4' = i.$$

Therefore, under these circumstances after the transformation at the corresponding spacetime point the velocity $\mathfrak{w}' = 0$, i.e. it is *transformed to rest* as we can express ourselves. We may call the invariant $\varrho\sqrt{1 - \mathfrak{w}^2}$ the *rest-density* of electricity.

§5.

Spacetime Vectors of the I^{st} and IInd kind

By taking into account the main result of the special Lorentz transformation together with the fact that the system of equations (A) as well the system of equations (B) is covariant with respect to a rotation of the reference system about the origin, we obtain the general *theorem of relativity*. To make the facts easily comprehensible, it may be more convenient to define a series of abbreviated terms for the purpose of expressing the ideas in a concise form, while on the other hand I shall adhere to the practice of using complex variables in order to make certain symmetries evident.

The linear homogeneous transformation

$$
\begin{aligned}
x_1 &= \alpha_{11}x_1' + \alpha_{12}x_2' + \alpha_{13}x_3' + \alpha_{14}x_4', \\
x_2 &= \alpha_{21}x_1' + \alpha_{22}x_2' + \alpha_{23}x_3' + \alpha_{24}x_4', \\
x_3 &= \alpha_{31}x_1' + \alpha_{32}x_2' + \alpha_{33}x_3' + \alpha_{34}x_4', \\
x_4 &= \alpha_{41}x_1' + \alpha_{42}x_2' + \alpha_{43}x_3' + \alpha_{44}x_4'
\end{aligned}
\tag{21}
$$

of determinant $+1$, in which all coefficients without the index 4 occurring once are real, while $\alpha_{14}, \alpha_{24}, \alpha_{34}$ and $\alpha_{41}, \alpha_{42}, \alpha_{43}$, are purely imaginary, but α_{44} is real and > 0, and through which

$$
x_1^2 + x_2^2 + x_3^2 + x_4^2 \quad \text{transforms into} \quad {x_1'}^2 + {x_2'}^2 + {x_3'}^2 + {x_4'}^2,
$$

I will call a general *Lorentz transformation*.

If we put

$$
x_1' = x', \quad x_2' = y', \quad x_3' = z', \quad x_4' = it',
$$

then it immediately arises from here a homogeneous linear transformation of x, y, z, t to x', y', z', t' with real coefficients, whereby the expression

$$
-x^2 - y^2 - z^2 + t^2 \quad \text{transforms into} \quad -{x'}^2 - {y'}^2 - {z'}^2 + {t'}^2,
$$

and to every such system of variables x, y, z, t with a *positive t*, for which this expression > 0, there always corresponds a *positive t'*; the latter is easily seen from the continuity of the expression in x, y, z, t.

The last vertical column of the system of coefficients (21) has to satisfy the condition

$$
\alpha_{14}^2 + \alpha_{24}^2 + \alpha_{34}^2 + \alpha_{44}^2 = 1.
\tag{22}
$$

If $\alpha_{14} = 0$, $\alpha_{24} = 0$, $\alpha_{34} = 0$, then $\alpha_{44} = 1$ and the Lorentz transformation reduces to a simple rotation of the spatial coordinate system about the origin.

If not all $\alpha_{14}, \alpha_{24}, \alpha_{34}$ are zero and we set

$$\alpha_{14} : \alpha_{24} : \alpha_{34} : \alpha_{44} = \mathfrak{b}_x : \mathfrak{b}_y : \mathfrak{b}_z : i,$$

then the absolute quantity

$$q = \sqrt{\mathfrak{b}_x^2 + \mathfrak{b}_y^2 + \mathfrak{b}_z^2} < 1.$$

follows from (22). On the other hand, for every set of coefficients α_{14}, α_{24}, α_{34}, α_{44} satisfying the condition (22) with real values of $\mathfrak{b}_x, \mathfrak{b}_y, \mathfrak{b}_z$, we can construct the *special* Lorentz transformation (6) with $\alpha_{14}, \alpha_{24}, \alpha_{34}, \alpha_{44}$ as the last vertical column, and then every Lorentz transformation with the same last vertical column of coefficients is composed of the special Lorentz transformation and a rotation of the spatial coordinate system about the origin.

The totality of all Lorentz transformations forms a *group*.

Under a *spacetime vector of the I-st kind* shall be understood a system of four quantities $\varrho_1, \varrho_2, \varrho_3, \varrho_4$ with the requirement that for every Lorentz transformation (21) it should be replaced by the system $\varrho_1', \varrho_2', \varrho_3', \varrho_4'$ which follows from (21) for the variables x_1', x_2', x_3', x_4' if for x_1, x_2, x_3, x_4 the values of $\varrho_1, \varrho_2, \varrho_3, \varrho_4$ are taken.

Besides the time-space vector of the I-st kind x_1, x_2, x_3, x_4 we shall also make use of another spacetime vector of the I-st kind u_1, u_2, u_3, u_4, and will consider the linear combination

$$\begin{aligned}
&f_{23}(x_2 u_3 - x_3 u_2) + f_{31}(x_3 u_1 - x_1 u_3) + f_{12}(x_1 u_2 - x_2 u_1) + \\
&+ f_{14}(x_1 u_4 - x_4 u_1) + f_{24}(x_2 u_4 - x_4 u_2) + f_{34}(x_3 u_4 - x_4 u_3)
\end{aligned} \tag{23}$$

with six coefficients $f_{23} \cdots f_{34}$. We note that in vectorial notation this can be constructed out of four vectors

$$x_1, x_2, x_3; \quad u_1, u_2, u_3; \quad f_{23}, f_{31}, f_{12}; \quad f_{14}, f_{24}, f_{34}$$

and the constants x_4 and u_4, and, on the other hand as a form symmetric in the indices $1, 2, 3, 4$. If we subject x_1, x_2, x_3, x_4 and u_1, u_2, u_3, u_4 simultaneously to the Lorentz transformation (21), the combination (23) is changed to

$$\begin{aligned}
&f_{23}'(x_2' u_3' - x_3' u_2') + f_{31}'(x_3' u_1' - x_1' u_3') + f_{12}'(x_1' u_2' - x_2' u_1') + \\
&+ f_{14}'(x_1' u_4' - x_4' u_1') + f_{24}'(x_2' u_4' - x_4' u_2') + f_{34}'(x_3' u_4' - x_4' u_3')
\end{aligned} \tag{24}$$

where the six coefficients $f'_{23} \cdots f'_{34}$ depend solely on the six $f_{23} \cdots f_{34}$ and the sixteen coefficients $\alpha_{11}, \alpha_{12}, \cdots \alpha_{44}$.

We shall define a *spacetime vector of the II-nd kind* as a system of six quantities $f_{23}, f_{31}, f_{12}, f_{14}, f_{24}, f_{34}$ with the condition that when subjected to a Lorentz transformation, it is changed to a new system $f'_{23}, f'_{31}, f'_{12}, f'_{14}, f'_{24}, f'_{34}$ which satisfies the connection between (23) and (24).

Now I express the general theorem of relativity concerning the equations (I) - (IV), which are the "fundamental equations for Aether," as follows.

If x, y, z, it (space coordinates, and time $t \times i$) is subjected to a Lorentz transformation, and at the same time $\varrho\mathfrak{w}_x, \varrho\mathfrak{w}_y, \varrho\mathfrak{w}_z, i\varrho$ (convection current, and charge density $\times i$) is transformed as a spacetime vector of the I-st kind, while $\mathfrak{m}_x, \mathfrak{m}_y, \mathfrak{m}_z, -i\mathfrak{e}_x, -i\mathfrak{e}_y, -i\mathfrak{e}_z$ (magnetic force, and electric induction $\times -i$) is transformed as a spacetime vector of the II-nd kind, then the system of equations (I), (II) and the system of equations (III), (IV) transform into exactly the same relationships between the respective newly introduced variables.

These facts can be more concisely expressed in these words: the system of equations (I), and (II) as well as the system of equations (III), (IV) are covariant under any Lorentz transformation, where $\varrho\mathfrak{w}, i\varrho$ should be transformed as a spacetime vector of the I-st kind, whereas $\mathfrak{m}, -i\mathfrak{e}$ should be treated as a vector of the II-nd kind. Or even more concisely:

$\varrho\mathfrak{w}, i\varrho$ *is a spacetime vector of the I-st kind,* $\mathfrak{m}, -i\mathfrak{e}$ *is a spacetime vector of the II-nd kind.*

I add a few comments here to clarify the idea of spacetime vector of the II-nd kind. *Invariants* of such a vector $\mathfrak{m}, -i\mathfrak{e}$ for the group of Lorentz transformations are obviously

$$\mathfrak{m}^2 - \mathfrak{e}^2 = f_{23}^2 + f_{31}^2 + f_{12}^2 + f_{14}^2 + f_{24}^2 + f_{34}^2, \tag{25}$$

$$\mathfrak{m}\mathfrak{e} = i(f_{23}f_{14} + f_{31}f_{24} + f_{12}f_{34}). \tag{26}$$

A spacetime vector of the II-nd kind $\mathfrak{m}, -i\mathfrak{e}$ (where $\mathfrak{m}$ and $\mathfrak{e}$ are real space vectors), may be called *singular*, when the scalar square $(\mathfrak{m} - i\mathfrak{e})^2 = 0$, i.e. $\mathfrak{m}^2 - \mathfrak{e}^2 = 0$, and at the same time $(\mathfrak{m}\mathfrak{e}) = 0$, i.e. *the vectors $\mathfrak{m}$ and $\mathfrak{e}$ are equal and perpendicular to each other*. If such is the case, these two properties for the spacetime vector of the II-nd kind remain unchanged for for every Lorentz transformation.

If the spacetime vector of the II-nd kind $\mathfrak{m}, -i\mathfrak{e}$ *is not singular*, we rotate the spacial coordinate system in such a way that the vector product $[\mathfrak{m}\mathfrak{e}]$ is in the direction of the $z-$axis, then $\mathfrak{m}_z = 0$, $\mathfrak{e}_z = 0$. Then

$$(\mathfrak{m}_x - i\mathfrak{e}_x)^2 + (\mathfrak{m}_y - i\mathfrak{e}_y)^2 \neq 0,$$

that is $\frac{\mathfrak{e}_y+i\mathfrak{m}_y}{\mathfrak{e}_x+i\mathfrak{m}_x}$ is different from $\pm i$ and therefore we can determine a complex argument $\varphi + i\psi$ in such a way that

$$\text{tg}(\varphi + i\psi) = \frac{\mathfrak{e}_y + i\mathfrak{m}_y}{\mathfrak{e}_x + i\mathfrak{m}_x}.$$

Then taking into account the equation (9), we combine the transformation (1) with a parameter ψ and a subsequent rotation about the $z-$axis at an angle φ, we perform a Lorentz transformation and at the end $\mathfrak{m}_y = 0$, $\mathfrak{e}_y = 0$, and therefore $\mathfrak{m}$ and $\mathfrak{e}$ will both be directed along the new $x-$axis; through the invariants $\mathfrak{m}^2 - \mathfrak{e}^2$ and $(\mathfrak{m}\mathfrak{e})$ one can fix the final values of these vectors, and also whether they are of the same or of opposite directions, or whether one of them is equal to zero.

§6.

Concept of Time

Due to the Lorentz transformations certain changes to the time parameter are allowed. Consequently, it is no longer permissible to speak of the *simultaneity* of two events *per se*. The use of this term implies, rather, that the freedom of choosing 6 parameters for defining a reference system for space and time is already limited in some way to freedom of choosing only 3 parameters. Only because we are used to this constraint, we consider the concept of simultaneity of two events as inherently existent[8]. The actual situation may be described in the following way.

Let a reference system x, y, z, t for spacetime points (events) be somehow known. If a space point A (x_0, y_0, z_0) at the time $t_0 = 0$ is compared with a space point P (x, y, z) at the time t and if the time difference $t - t_0$ (let $t > t_0$) is *smaller* than the length AP i.e. smaller than the time required for the propagation of light from A to P, and if q is the ration $\frac{t-t_0}{AP} < 1$, then by a special Lorentz transformation, in which AP is taken as the axis, and which has the moment q, we can introduce a new time parameter t' which (see equation (12) in §4) has the same value $t' = 0$ for both spacetime points A, t_0, and P, t; therefore the two events can be regarded as simultaneous.

Let us take at the same time $t_0 = 0$, two different space-points A, B, or three space-points (A, B, C) which do not lie on the same space-line, and

---

[8] Just as beings which are confined within a narrow region surrounding a point on a spherical surface may be led to think that a sphere is a geometric figure in which one diameter is specifically distinguished from the others.

compare with them a space point P, which is outside the line AB, or the plane ABC at another time t, and let the time difference $t - t_0$ $(t > t_0)$ be smaller than the time which light requires for propagation from the line AB, or the plane ABC to P. Let q be the quotient of the first to the second time; then if a Lorentz transformation is carried out in which the perpendicular from P on AB, or from P on the plane ABC is the axis, and q is the moment, then all the three (or four) events A, t_0; B, t_0; (C, t_0) and P, t are simultaneous.

If four space-points, which do not lie on one plane are conceived to be at the same time t_0, then it is no longer possible to make a change of the time parameter by a Lorentz transformation without at the same time destroying the character of the simultaneity of these four space points.

To the mathematician, accustomed on the one hand to the above considerations in multidimensional manifolds, and on the other hand to the concepts of the so-called non-Euclidean geometry, there can be no difficulty in adopting this concept of time for the application of the Lorentz transformation. The paper of Einstein which has been cited in the Introduction, has succeeded to some extent in presenting the nature of the transformation from a physical standpoint.

Second Part.
The electromagnetic phenomena

§7.

Fundamental Equations for Bodies at Rest

After these preparatory works, which have been first developed on account of the small amount of mathematics involved in the limiting case $\varepsilon = 1, \mu = 1, \sigma = 0$, let us turn to the electromagnetic phenomena in matter. We look for those relations which make it possible for us – under appropriate boundary condition – to obtain the following quantities at every place and time, and therefore at every spacetime point as functions of x, y, z, t: the vector of the electric force $\mathfrak{E}$, the magnetic induction $\mathfrak{M}$, the electrical induction $\mathfrak{e}$, the magnetic force $\mathfrak{m}$, the electrical space-density ϱ, the vector of the "electric current" $\mathfrak{s}$ (whose relation hereafter to the conduction current is known by the manner in which conductivity occurs in the process), and lastly the vector $\mathfrak{w}$, the velocity of matter.

The relations in question can be divided into two classes,

firstly, those equations, which by using the vector $\mathfrak{w}$, the velocity of matter, as a function of x, y, z, t, allow us to determine other quantities as functions of x, y, z, t I will specifically call this first class of equations *the fundamental equations*,

secondly, the expressions for the *ponderomotive forces*, which by using the laws of mechanics give us additional information about the vector $\mathfrak{w}$ as a function of x, y, z, t.

For the case of a body at rest, i.e. when $\mathfrak{w}(x, y, z, t) = 0$, the theories of Maxwell (Heaviside, Hertz) and Lorentz lead to the same fundamental equations. These are

1) the *differential equations*, which contain no constants related to matter:

$$\operatorname{curl} \mathfrak{m} - \frac{\partial \mathfrak{e}}{\partial t} = \mathfrak{s}, \tag{I}$$

$$\operatorname{div} \mathfrak{e} = \varrho, \tag{II}$$

$$\operatorname{curl} \mathfrak{E} + \frac{\partial \mathfrak{M}}{\partial t} = 0, \tag{III}$$

$$\operatorname{div} \mathfrak{M} = 0. \tag{IV}$$

2) additional relationships, which characterize the influence of existing matter; for the most important case, to which we limit ourselves, i.e. for isotopic bodies, they take the form

$$\mathfrak{e} = \varepsilon \mathfrak{E} = 0, \qquad \mathfrak{M} = \mu \mathfrak{m}, \qquad \mathfrak{s} = \sigma \mathfrak{E}, \tag{V}$$

where the dielectric constant ε, the magnetic permeability μ, the conductivity of matter σ, are all regarded as known function of x, y, z and t. Here $\mathfrak{s}$ is the *conduction current*.

I will now return to these equations by a modified notation and an even more hidden symmetry will emerge. I use as in the above text

$$x_1 = x, \quad x_2 = y, \quad x_3 = z, \quad x_4 = it$$

and will write

$$s_1, \quad s_2, \quad s_3, \quad s_4$$

instead of

$$\mathfrak{s}_x, \quad \mathfrak{s}_y, \quad \mathfrak{s}_z, \quad i\varrho,$$

further

$$f_{23}, \quad f_{31}, \quad f_{12}, \quad f_{14}, \quad f_{24}, \quad f_{34}$$

for

$$\mathfrak{m}_x, \quad \mathfrak{m}_y, \quad \mathfrak{m}_z, \quad -i\mathfrak{e}_x, \quad -i\mathfrak{e}_y, \quad -i\mathfrak{e}_z,$$

and also

$$F_{23}, \quad F_{31}, \quad F_{12}, \quad F_{14}, \quad F_{24}, \quad F_{34}$$

for

$$\mathfrak{M}_x, \quad \mathfrak{M}_y, \quad \mathfrak{M}_z, \quad -i\mathfrak{E}_x, \quad -i\mathfrak{E}_y, \quad -i\mathfrak{E}_z;$$

finally for all unequal pairs h, k of indices $1, 2, 3, 4$ the relation

$$f_{kh} = -f_{hk}, \qquad F_{kh} = -F_{hk},$$

should be satisfied. (The letters f, F should remind us of the word field (Feld), whereas s – the word current (Strom).)

Then the fundamental equations (I), (II) can be written as:

$$\frac{\partial f_{12}}{\partial x_2} + \frac{\partial f_{13}}{\partial x_3} + \frac{\partial f_{14}}{\partial x_4} = s_1,$$

$$\frac{\partial f_{21}}{\partial x_1} + \frac{\partial f_{23}}{\partial x_3} + \frac{\partial f_{24}}{\partial x_4} = s_2,$$

$$\frac{\partial f_{31}}{\partial x_1} + \frac{\partial f_{32}}{\partial x_2} + \frac{\partial f_{34}}{\partial x_4} = s_3, \qquad \text{(A)}$$

$$\frac{\partial f_{41}}{\partial x_1} + \frac{\partial f_{42}}{\partial x_2} + \frac{\partial f_{43}}{\partial x_3} = s_4,$$

and the equations (III) and (IV) as

$$\frac{\partial F_{34}}{\partial x_2} + \frac{\partial F_{42}}{\partial x_3} + \frac{\partial F_{23}}{\partial x_4} = 0,$$

$$\frac{\partial F_{43}}{\partial x_1} + \frac{\partial F_{14}}{\partial x_3} + \frac{\partial F_{31}}{\partial x_4} = 0,$$

$$\frac{\partial F_{24}}{\partial x_1} + \frac{\partial F_{41}}{\partial x_2} + \frac{\partial F_{12}}{\partial x_4} = 0, \qquad \text{(B)}$$

$$\frac{\partial F_{32}}{\partial x_1} + \frac{\partial F_{13}}{\partial x_2} + \frac{\partial F_{21}}{\partial x_3} = 0.$$

§8.

The Fundamental Equations for Moving Bodies

We can now determine in an unambiguous way the fundamental equations for arbitrarily moving bodies through the following three axioms:

The *first* axiom should be:

When a detached region of matter is at rest at any moment, so in a system x, y, z, t the vector $\mathfrak{w}$ is zero – the surroundings of that region may be understood to be in any motion – then for the spacetime point x, y, z, t the same relations (A), (B), (V), which hold when all matter is at rest, should also hold between ϱ, and the vectors $\mathfrak{s}$, $\mathfrak{e}$, $\mathfrak{m}$, $\mathfrak{E}$, $\mathfrak{M}$ and their derivatives with respect to x, y, z, t.

The *second* axiom should be:

Every velocity of matter is < 1, i.e. smaller than the velocity of propagation of light.

The *third* axiom should be:

The fundamental equations are of such kind that when x, y, z, it are subjected to a Lorentz transformation and the the pairs $\mathfrak{m}, -i\mathfrak{e}$ and $\mathfrak{M}, -i\mathfrak{E}$ are transformed as spacetime vectors of the II-nd kind, whereas $\mathfrak{s}, i\varrho$ is transformed as a spacetime vector of the I-st kind, the equations are transformed into essentially identical forms involving the transformed quantities.

I will express this third axiom shortly as:

$\mathfrak{m}, -i\mathfrak{e}$ and $\mathfrak{M}, -i\mathfrak{E}$ *are spacetime vectors of the II-nd kind, whereas $\mathfrak{s}, i\varrho$ is a spacetime vector of the I-st kind.* I call this axiom the *principle of relativity*.

In fact these three axioms unambiguously lead us from the previously formulated fundamental equations for bodies at rest to the fundamental equations for moving bodies.

Namely, according to the second axiom in each spacetime point the magnitude of the velocity vector $|\mathfrak{w}| < 1$. As a result, we can always represent the vector $\mathfrak{w}$ by the following set of four quantities

$$w_1 = \frac{\mathfrak{w}_x}{\sqrt{1 - \mathfrak{w}^2}}, \qquad w_2 = \frac{\mathfrak{w}_y}{\sqrt{1 - \mathfrak{w}^2}},$$

$$w_3 = \frac{\mathfrak{w}_z}{\sqrt{1 - \mathfrak{w}^2}}, \qquad w_4 = \frac{i}{\sqrt{1 - \mathfrak{w}^2}},$$

between which the relationship

$$w_1^2 + w_2^2 + w_3^2 + w_4^2 = -1 \tag{27}$$

holds.

From the concluding remarks of §4 it is seen that under Lorentz transformations this quadruple behaves as a spacetime vector of the I-st kind, and we will to call it *spacetime velocity vector*.

Let us now have a specific point x, y, z of matter at a given time t. If at this spacetime point $\mathfrak{w} = 0$, then, according to the first axiom, we have for this point the equations (A), (B), (V) of §7. If $\mathfrak{w} \neq 0$ and since $|\mathfrak{w}| < 0$ then there exists according to (16) a special Lorentz transformation, whose vector $\mathfrak{b}$ is equal to the vector $\mathfrak{w}(x, y, z, t)$ and this transformation leads us to a new system of reference x', y', z', t'. For the considered spacetime as we saw in §4 new values arise

$$w_1' = 0, \quad w_2' = 0, \quad w_3' = 0, \quad w_4' = i, \tag{28}$$

so the new velocity vector $\mathfrak{w}' = 0$ and *the spacetime point is*, as we put it then, *as if transformed to rest*. Now according to the third axiom from the fundamental equations for the spacetime point x, y, z, t one should obtain the corresponding fundamental equations for the point x', y', z', t', expressed through the transformed quantities $\mathfrak{w}', \varrho', \mathfrak{s}', \mathfrak{e}', \mathfrak{m}', \mathfrak{E}', \mathfrak{M}'$ and their derivatives with respect to x', y', z', t'. According to the first axiom and since $\mathfrak{w}' = 0$ these equations must be exactly equivalent to

1) the differential equations (A'), (B') which are obtained from the equations (A), (B) by simply priming the symbols in (A) and (B),

2) and the equations

$$\varrho' = \varepsilon \mathfrak{E}' \quad , \mathfrak{M}' = \mu \mathfrak{m}' \quad , \mathfrak{s}' = \sigma \mathfrak{E}', \tag{V'}$$

where ε, μ, σ are the dielectric constant, magnetic permeability, and conductivity for the system x', y', z', t' corresponding to the properties of matter in the spacetime point x, y, z, t.

Now we go through the inverse Lorentz transformation to the original variables x, y, z, t, and the quantities $\mathfrak{w}, \varrho, \mathfrak{s}, \mathfrak{e}, \mathfrak{m}, \mathfrak{E}, \mathfrak{M}$; then the equations, which we obtain from this transformation, will be the fundamental equations sought by us for the moving bodies.

Now from §4, and §6, it can be seen that the system of equations (A), as well as the system of equations (B) are covariant under the Lorentz transformations, i.e. the equations, which we obtain backwards from (A') (B'), must be exactly of the same form as the equations (A) and (B), which hold for bodies at rest. We have therefore as the first result:

The differential equations expressing the fundamental equations of electrodynamics for moving bodies, when written in ϱ and the vectors $\mathfrak{s}, \mathfrak{e}, \mathfrak{m}, \mathfrak{E}$,

$\mathfrak{M}$ *are exactly of the same form as the equations for bodies at rest.* The velocity of matter does not enter in these equations. In vector form these equations are therefore once again

$$\text{curl } \mathfrak{m} - \frac{\partial \mathfrak{e}}{\partial t} = \mathfrak{s}, \tag{I}$$

$$\text{div } \mathfrak{e} = \varrho, \tag{II}$$

$$\text{curl } \mathfrak{E} + \frac{\partial \mathfrak{M}}{\partial t} = 0, \tag{III}$$

$$\text{div } \mathfrak{M} = 0. \tag{IV}$$

The velocity of the matter is only referred to the additional conditions that characterize the influence of matter through the special constants ε, μ, σ. Let us now transform these additional conditions (V') into the original coordinates x, y, z, and the original time t.

According to formula (15) in §4, the component of $\mathfrak{e}'$ in the direction of the vector $\mathfrak{w}$ is the same us that of $\mathfrak{e} + [\mathfrak{w}\mathfrak{m}]$, the component of $\mathfrak{m}'$ is the same as that of $\mathfrak{m} - [\mathfrak{w}\mathfrak{e}]$, but for the perpendicular direction $\overline{\mathfrak{w}}$ the components of $\mathfrak{e}'$ and $\mathfrak{m}'$ are the same as those of $\mathfrak{e} + [\mathfrak{w}\mathfrak{m}]$ and $\mathfrak{m} - [\mathfrak{w}\mathfrak{e}]$, multiplied by $\frac{1}{\sqrt{1-\mathfrak{w}^2}}$. On the other hand, $\mathfrak{E}'$ and $\mathfrak{M}'$ stand to $\mathfrak{E} + [\mathfrak{w}\mathfrak{M}]$, and $\mathfrak{M} - [\mathfrak{w}\mathfrak{E}]$ in the same relation as $\mathfrak{e}'$ and $\mathfrak{m}'$ to $\mathfrak{e} + [\mathfrak{w}\mathfrak{m}]$ and $\mathfrak{m} - [\mathfrak{w}\mathfrak{e}]$. Then, by distinguishing in the relation $\mathfrak{e}' = \varepsilon \mathfrak{E}'$ three components – one parallel to the vector $\mathfrak{w}$ and the other two mutually perpendicular and perpendicular to $\mathfrak{w}$ – in which case the last two components are multiplied by $\sqrt{1 - \mathfrak{w}^2}$, we have

$$\mathfrak{e} + [\mathfrak{w}\mathfrak{m}] = \varepsilon(\mathfrak{E} + [\mathfrak{w}\mathfrak{M}]). \tag{C}$$

Analogously, the relation $\mathfrak{M}' = \mu \mathfrak{m}'$ leads to

$$\mathfrak{M} - [\mathfrak{w}\mathfrak{E}] = \mu(\mathfrak{m} - [\mathfrak{w}\mathfrak{e}]). \tag{D}$$

Further it follows from the transformation equations (12), (10), (11) in §4, when we replace $q, \mathfrak{r}_{\mathfrak{b}}, \mathfrak{r}_{\overline{\mathfrak{b}}}, t, \mathfrak{r}'_{\mathfrak{b}}, \mathfrak{r}'_{\overline{\mathfrak{b}}}, t'$ by $|\mathfrak{w}|, \mathfrak{s}_{\mathfrak{w}}, \mathfrak{s}_{\overline{\mathfrak{w}}}, \varrho, \mathfrak{s}'_{\mathfrak{w}}, \mathfrak{s}'_{\overline{\mathfrak{w}}}, \varrho'$ that

$$\varrho' = \frac{-|\mathfrak{w}|\mathfrak{s}_{\mathfrak{w}} + \varrho}{\sqrt{1 - \mathfrak{w}^2}}, \quad \mathfrak{s}'_{\mathfrak{w}} = \frac{\mathfrak{s}_{\mathfrak{w}} - |\mathfrak{w}|\varrho}{\sqrt{1 - \mathfrak{w}^2}}, \quad \mathfrak{s}'_{\overline{\mathfrak{w}}} = \mathfrak{s}_{\overline{\mathfrak{w}}}$$

so that from $\mathfrak{s}' = \sigma \mathfrak{E}'$ now we have

$$\frac{\mathfrak{s}_{\mathfrak{w}} - |\mathfrak{w}|\varrho}{\sqrt{1 - \mathfrak{w}^2}} = \sigma(\mathfrak{E} + [\mathfrak{w}\mathfrak{M}])_{\mathfrak{w}}$$

$$\mathfrak{s}_{\overline{\mathfrak{w}}} = \frac{\sigma(\mathfrak{E} + [\mathfrak{w}\mathfrak{M}])_{\overline{\mathfrak{w}}}}{\sqrt{1 - \mathfrak{w}^2}} \qquad\qquad (E)$$

According to the way in which the conductivity σ enters into these relations it will be appropriate to call the vector $\mathfrak{s} - \varrho\mathfrak{w}$ *conduction current* with components $\mathfrak{s}_{\mathfrak{w}} - \varrho|\mathfrak{w}|$ in the direction of $\mathfrak{w}$, and $\mathfrak{s}_{\overline{\mathfrak{w}}}$ in the direction of $\overline{\mathfrak{w}}$; this vector vanishes for $\sigma = 0$.

We remark that for $\varepsilon = 1$, $\mu = 1$ the equations $\mathfrak{e}' = \mathfrak{E}'$, $\mathfrak{m}' = \mathfrak{M}'$ through the inverse Lorentz transformation with the vector $-\mathfrak{w}$ are immediately transformed to the equations $\mathfrak{e} = \mathfrak{E}$, $\mathfrak{m} = \mathfrak{M}$, and for $\sigma = 0$ the equation $\mathfrak{s}' = 0$ becomes $\mathfrak{s} = \varrho\mathfrak{w}$, so that as a result the limiting case of the equations obtained here with $\varrho = 1$, $\mu = 1$, $\sigma = 0$ turn out to be the "fundamental equations of Aether" discussed in §2.

§9.

The fundamental equations in the theory of Lorentz

Now let us see to what extent the fundamental equations accepted by Lorentz correspond to the relativity postulate as defined in §8. In the article "Electron theory" (Enzykl. der math. Wiss., V2, Art. 14) Lorentz used differential equations for any possible, even magnetized bodies (see there page 209, Eq. XXX', formula (14) on page 78 of the same issue):

$$\operatorname{curl}(\mathfrak{H} - [\mathfrak{w}\mathfrak{E}]) = \mathfrak{J} + \frac{\partial \mathfrak{D}}{\partial t} + \mathfrak{w}\operatorname{div}\mathfrak{D} - \operatorname{curl}[\mathfrak{w}\mathfrak{D}] \qquad (IIIa'')$$

$$\operatorname{div}\mathfrak{D} = \varrho \qquad (I'')$$

$$\operatorname{curl}\mathfrak{E} = -\frac{\partial \mathfrak{B}}{\partial t} \qquad (IV'')$$

$$\operatorname{div}\mathfrak{B} = 0. \qquad (IV'')$$

Then for moving non-magnetized bodies, Lorentz puts (page 223, 3) $\mu = 1$, $\mathfrak{B} = \mathfrak{H}$ and assumes the occurrence of the dielectric constant ε and conductivity σ according to

$$\mathfrak{D} - \mathfrak{E} = (\varepsilon - 1)(\mathfrak{E} + [\mathfrak{w}\mathfrak{E}]), \qquad \text{(Eq. XXXIV''', p. 227)}$$

$$\mathfrak{J} = \sigma(\mathfrak{E} + [\mathfrak{w}\mathfrak{E}]). \qquad \text{(Eq. XXXIII'', p. 223)}$$

The Lorentz symbols $\mathfrak{E}, \mathfrak{B}, \mathfrak{H}, \mathfrak{D}$ are replaced here by $\mathfrak{E}, \mathfrak{M}, \mathfrak{e}, \mathfrak{m}$ whereas Lorentz denotes the conduction current by $\mathfrak{J}$.

The last three of the differential equations cited above now coincide immediately with the equations (II), (III), (IV) here, but the first equation, since $\sigma = 0$ for vanishing currents the vector $\mathfrak{J}$ can be identified with the current $\mathfrak{s} - \mathfrak{w}\varrho$, becomes

$$\operatorname{curl}\left(\mathfrak{H} - [\mathfrak{w}\mathfrak{E}]\right) = \mathfrak{s} + \frac{\partial \mathfrak{D}}{\partial t} - \operatorname{curl}[\mathfrak{w}\mathfrak{D}], \tag{29}$$

which does not coincide with (I) here. Therefore the general differential equations of Lorentz for any magnetized body do *not* correspond to the relativity principle.

On the other hand, the form corresponding to the relativity principle for the condition of non-magnetization should be taken from (D) in §8 with $\mu = 1$ not as $\mathfrak{B} = \mathfrak{H}$, as Lorentz did, but as

$$\mathfrak{B} - [\mathfrak{w}\mathfrak{E}] \qquad (\text{from} \quad \mathfrak{M} - [\mathfrak{w}\mathfrak{E}] = \mathfrak{m} - [\mathfrak{w}\mathfrak{e}]), \tag{30}$$

Now by putting $\mathfrak{H} = \mathfrak{B}$, the differential equation (29) is transformed into the same form as equation (I) here when $\mathfrak{m} - [\mathfrak{w}\mathfrak{e}] = \mathfrak{M} - [\mathfrak{w}\mathfrak{E}]$. So it comes with a compensation of two contradictions with the relativity principle that the differential equations of Lorentz for moving non-magnetized bodies finally agree with the relativity principle.

If we make use of (30) for non-magnetized bodies, and put accordingly $\mathfrak{H} = \mathfrak{B} + [\mathfrak{w}, \mathfrak{D} - \mathfrak{E}]$, then as a consequence of (C) in §8

$$(\varepsilon - 1)(\mathfrak{E} + [\mathfrak{w}\mathfrak{B}]) = \mathfrak{D} - \mathfrak{E} + [\mathfrak{w}[\mathfrak{w}, \mathfrak{D} - \mathfrak{E}]],$$

i.e. in the direction of $\mathfrak{w}$

$$(\varepsilon - 1)(\mathfrak{E} + [\mathfrak{w}\mathfrak{B}])_\mathfrak{w} = (\mathfrak{D} - \mathfrak{E})_\mathfrak{w},$$

and in a direction $\overline{\mathfrak{w}}$ perpendicular to $\mathfrak{w}$

$$(\varepsilon - 1)(\mathfrak{E} + [\mathfrak{w}\mathfrak{B}])_{\overline{\mathfrak{w}}} = (1 - \mathfrak{w}^2)(\mathfrak{D} - \mathfrak{E})_{\overline{\mathfrak{w}}},$$

i.e. it coincides with Lorentz' assumption, if we neglect terms of the order of $\mathfrak{w}^2$ which are small compared to 1.

Also to the same order of approximation, Lorentz' form for $\mathfrak{J}$ corresponds to the conditions imposed by the relativity principle (comp. (E) in §8) that the components $\mathfrak{J}_\mathfrak{w}$ and $\mathfrak{J}_{\overline{\mathfrak{w}}}$ should be equal to the components of $\sigma(\mathfrak{E} + \mathfrak{w}\mathfrak{B})$ multiplied by $\sqrt{1 - \mathfrak{w}^2}$ and $\frac{1}{\sqrt{1-\mathfrak{w}^2}}$, respectively.

§10.

The fundamental equations of E. Cohn

E. Gohn[9] adopted the following fundamental equations:

$$\operatorname{curl}\left(M + [\mathfrak{w}\mathfrak{E}]\right) = \frac{\partial \mathfrak{E}}{\partial t} + \mathfrak{w}\operatorname{div}\mathfrak{E} + \mathfrak{J}$$

$$-\operatorname{curl}\left(E - [\mathfrak{w}\mathfrak{M}]\right) = \frac{\partial \mathfrak{M}}{\partial t} + \mathfrak{w}\operatorname{div}\mathfrak{M},$$

$$(31)$$

$$\mathfrak{J} = \sigma E, \quad \mathfrak{E} = \varepsilon E - [\mathfrak{w}M]), \quad \mathfrak{M} = \mu M + [\mathfrak{w}E], \tag{32}$$

where where E, M are the electric and magnetic field intensities (forces), $\mathfrak{E}, \mathfrak{M}$ are the electric and magnetic polarization (induction). These equations also permit the existence of true magnetism; if we do not take into account this consideration, it should be taken that div $\mathfrak{M} = 0$.

An objection to this system of equations is that for $\varepsilon = 1, \mu = 1$ the vectors force and induction do not coincide. If, however, in the equations for the electric and magnetic force we use not E and M, but $E - [\mathfrak{w}\mathfrak{M}]$ and $M + [\mathfrak{w}\mathfrak{E}]$, and substitute $\mathfrak{E}, \mathfrak{M}, E, M, \operatorname{div}\mathfrak{E}$ with $\mathfrak{e}, \mathfrak{M}, \mathfrak{E} + [\mathfrak{w}\mathfrak{M}], \mathfrak{m} - [\mathfrak{w}\mathfrak{e}], \varrho$, then his differential equations transform to our equations, and the conditions (32) take the form

$$\mathfrak{J} = \sigma(\mathfrak{E} + [\mathfrak{w}\mathfrak{M}]),$$

$$\mathfrak{e} + [\mathfrak{w}, \mathfrak{m} - [\mathfrak{w}\mathfrak{e}]] = \varepsilon(\mathfrak{E} + [\mathfrak{w}\mathfrak{M}]),$$

$$\mathfrak{M} - [\mathfrak{w}, \mathfrak{E} + [\mathfrak{w}\mathfrak{M}]] = \mu(\mathfrak{m} - [\mathfrak{w}\mathfrak{e}]),$$

and then the equations of Cohn obtain the form required by the relativity principle, if terms of the order of $\mathfrak{w}^2$ are neglected compared to 1.

It should be mentioned that the equations of Hertz (in the notation of Cohn) differ from (31) with the additional conditions

$$\mathfrak{E} = \varepsilon E, \quad \mathfrak{M} = \mu M, \quad \mathfrak{J} = \sigma E, \tag{33}$$

---

[9]Gött. Nachr. 1901, S. 74 (also in Ann. d. Phys. 7 (4), 1902, S. 29)

but it seems it is not possible to interpret the symbols there in terms of observable quantities in order to arrive at an agreement with the relativity principle even as an approximation of the order of $\mathfrak{w}^2$.

§11.

Typical representation of the fundamental equations

In the statement of the fundamental equations, our leading idea had been that they should retain a covariance of form, when subjected to a group of Lorentz transformations. Now we have to deal with ponderomotive reactions and energy in the electromagnetic field and there should be no doubt that the resolution of this question is in some way connected with the simplest forms which can be given to the fundamental equations, satisfying the conditions of covariance. In order to arrive at such forms, I shall first of all put the fundamental equations in a *typical form which brings out clearly their covariance in case of a Lorentz transformation.* Here I am using a method of calculation, which enables us to deal in a simple manner with the spacetime vectors of the I-st, and II-nd kind, and of which the rules, as far as required are given below.

1^0. A system of quantities

$$\begin{vmatrix} a_{11}, & \cdots, & a_{1q} \\ \vdots & & \vdots \\ a_{p1}, & \ldots, & a_{pq} \end{vmatrix},$$

arranged in p horizontal rows, and q vertical columns is called a $p \times q$ *matrix*[10] and will be denoted by the letter A.

If all the quantities a_{hk} are multiplied by c, the resulting matrix with quantities ca_{hk} will be denoted by cA.

If the roles of the horizontal rows and vertical columns be interchanged, we obtain a $q \times p$ matrix, which will be known as the *transposed* matrix of A, and will be denoted by $\overline{A}$:

$$\overline{A} = \begin{vmatrix} a_{11}, & \cdots, & a_{p1} \\ \vdots & & \vdots \\ a_{1q}, & \ldots, & a_{pq} \end{vmatrix}.$$

---

[10]You might also think that instead of Cayley's matrix calculus one could use Hamilton's quaternion calculus, but the latter seems to me to be too restrictive and cumbersome for our purposes.

If we have a second matrix with the same p and q series as in A

$$B = \begin{vmatrix} b_{11}, & \cdots, & a_{1q} \\ \vdots & & \vdots \\ b_{p1}, & \cdots, & a_{pq} \end{vmatrix},$$

then $A + B$ shall denote the same $p \times q$ matrix whose members are $a_{hk} + b_{hk}$.

2^0. If we have two matrices

$$A = \begin{vmatrix} a_{11}, & \cdots, & a_{1q} \\ \vdots & & \vdots \\ a_{p1}, & \cdots, & a_{pq} \end{vmatrix}, \qquad B = \begin{vmatrix} b_{11}, & \cdots, & a_{1r} \\ \vdots & & \vdots \\ b_{q1}, & \cdots, & a_{qr} \end{vmatrix},$$

where *the number of horizontal rows of B, is equal to the number of vertical columns of A*, then by AB, the *product* of the matrices A and B, will be denoted the matrix

$$C = \begin{vmatrix} c_{11}, & \cdots, & c_{1r} \\ \vdots & & \vdots \\ c_{p1}, & \cdots, & c_{qr} \end{vmatrix},$$

where each element is formed by combination of the horizontal rows of A with the vertical columns of B according to the rule

$$c_{hk} = a_{h1}b_{1k} + a_{h2}b_{2k} + \ldots + a_{hk}b_{qk} \qquad \left(\begin{matrix} h = 1, 2, \ldots, p \\ k = 1, 2, \ldots, r \end{matrix} \right)$$

For such products, the *associative* law $(AB)S = A(BS)$ holds, where S is a third matrix whose horizontal rows are equal to the vertical columns of B (or AB).

For the transposed matrix of $C = BA$ we have $\overline{C} = \overline{BA}$.

3^0. Here it will only be considered matrices with no more than four horizontal rows and no more than four vertical rows.

A *unit matrix* (in matrix equations it will be for brevity given as 1) will be defined as the following 4×4 matrix with the elements

$$\begin{vmatrix} e_{11}, & e_{12}, & e_{13}, & e_{14} \\ e_{21}, & e_{22}, & e_{23}, & e_{24} \\ e_{31}, & e_{32}, & e_{33}, & e_{34} \\ e_{41}, & e_{42}, & e_{43}, & e_{44} \end{vmatrix} = \begin{vmatrix} 1, & 0, & 0, & 0 \\ 0, & 1, & 0, & 0 \\ 0, & 0, & 1, & 0 \\ 0, & 0, & 0, & 1 \end{vmatrix} \qquad (34)$$

For products $c \cdot 1$ with the participation of the unit matrix (as a special case of the matrix product cA introduced in 1^0) for the sake of brevity we will simply write c in matrix equations.

For a 4×4 A matrix, $\operatorname{Det} A$ will denote the determinant formed of the 4×4 elements of the matrix. If $\operatorname{Det} A \neq 0$, then there is a *reciprocal* matrix A^{-1} corresponding to A so that $A^{-1}A = 1$.

A matrix

$$
f = \begin{vmatrix}
0, & f_{12}, & f_{13}, & f_{14} \\
f_{21}, & 0, & f_{23}, & f_{24} \\
f_{31}, & f_{32}, & 0, & f_{34} \\
f_{41}, & f_{42}, & f_{43}, & 0
\end{vmatrix},
$$

whose elements fulfil the relation $f_{kh} = -f_{hk}$, is called an *alternating* matrix. These relations say that the transposed matrix $\overline{f} = -f$. Then by f^* will be the dual, alternating matrix

$$
f^* = \begin{vmatrix}
0, & f_{34}, & f_{42}, & f_{23} \\
f_{43}, & 0, & f_{14}, & f_{31} \\
f_{24}, & f_{41}, & 0, & f_{12} \\
f_{32}, & f_{13}, & f_{21}, & 0
\end{vmatrix}. \tag{35}
$$

Then

$$
f^* f = f_{32} f_{14} + f_{13} f_{24} + f_{21} f_{34} \tag{36}
$$

is a 4×4 matrix in which all elements outside the main diagonal from top left to bottom right are zero and that all elements in this diagonal correspond to each other and are equal to the combination of the coefficients f given by (36). The determinant of f turns out to be the square of this combination and we want by $\operatorname{Det}^{\frac{1}{2}} f$ to denote the expression

$$
\operatorname{Det}^{\frac{1}{2}} f = f_{32} f_{14} + f_{13} f_{24} + f_{21} f_{34} \tag{37}
$$

4^0. Any linear transformation

$$
x_k = \alpha_{h1} x_1' + \alpha_{h2} x_2' + \alpha_{h3} x_3' + \alpha_{h4} x_4' \qquad (h = 1, 2, 3, 4) \tag{38}
$$

can be represented by the 4×4 matrix with coefficients

$$A = \begin{vmatrix} \alpha_{11}, & \alpha_{12}, & \alpha_{13}, & \alpha_{14} \\ \alpha_{21}, & \alpha_{22}, & \alpha_{23}, & \alpha_{24} \\ \alpha_{31}, & \alpha_{32}, & \alpha_{33}, & \alpha_{34} \\ \alpha_{41}, & \alpha_{42}, & \alpha_{43}, & \alpha_{44} \end{vmatrix}.$$

which we will call transformation A. By the transformation A, the expression

$$x_1^2 + x_2^2 + x_3^2 + x_4^2$$

changes into the quadratic form

$$\Sigma a_{hk} x_h' x_k' \qquad (h, k = 1, 2, 3, 4),$$

where

$$a_{hk} = \alpha_{1h}\alpha_{1k} + \alpha_{2h}\alpha_{2k} + \alpha_{3h}\alpha_{3k} + \alpha_{4h}\alpha_{4k},$$

so that the 4×4 (symmetric) matrix formed by the coefficients of this form turns out to be the product $\overline{A}A$ of the transposed matrix of A and the matrix A itself. If the transformation results in the form

$$x_1'^2 + x_2'^2 + x_3'^2 + x_4'^2,$$

the product of $\overline{A}$ and A must be the matrix 1.

$$\overline{A}A = 1. \tag{39}$$

This relation is satisfied by A if its elements correspond to a Lorentz transformation in the transformation (38). For the determinant of A it follows out from (39) that $(\text{Det } A)^2 = 1$, $\text{Det } A = \pm 1$. The condition (39) can be written as

$$A^{-1} = \overline{A}, \tag{40}$$

i.e. the reciprocal matrix of A is equivalent to the transposed matrix of A.

For A representing a Lorentz transformation, we find that $\text{Det } A = +1$, that the coefficients $\alpha_{14}, \alpha_{24}, \alpha_{34}, \alpha_{41}, \alpha_{42}, \alpha_{43}$ are purely imaginary (or zero), the other coefficients are real, and $\alpha_{43} > 0$.

5^0. Any spacetime vector of the I-st kind $s = (s_1, s_2, s_3, s_4)$ should be represented by a 1×4 matrix which is formed by its four *components*:

$$s = |s_1, s_2, s_3, s_4|, \tag{41}$$

then a Lorentz transformation A turns it into sA

A spacetime vector of the II-nd kind with components $f_{23}, f_{31}, f_{12}, f_{14}, f_{24}, f_{34}$ will be represented by the alternating matrix

$$f = \begin{vmatrix} 0, & f_{12}, & f_{13}, & f_{14} \\ f_{21}, & 0, & f_{23}, & f_{24} \\ f_{31}, & f_{32}, & 0, & f_{34} \\ f_{41}, & f_{42}, & f_{43}, & 0 \end{vmatrix}, \tag{42}$$

and under a Lorentz transformation (see the rules (23), (24) in §5) f should be replaced by $\overline{A} f A = A^{-1} f A$. Using the expression (37), we have the identity $\operatorname{Det}^{\frac{1}{2}} (\overline{A} f A) = \operatorname{Det} A \operatorname{Det}^{\frac{1}{2}} f$. Therefore $\operatorname{Det}^{\frac{1}{2}} f$ is an invariant under the Lorentz transformations (see also equation (26) in §5).

For the dual matrix f^* we have by taking into account (36):

$$(A^{-1} f^* A)(A^{-1} f A) = A^{-1} f^* f A = \operatorname{Det}^{\frac{1}{2}} f \cdot A^{-1} A = \operatorname{Det}^{\frac{1}{2}} f,$$

from where it is easy to see that the dual matrix f^* is a spacetime vector of the II-nd kind since it transforms exactly like the spacetime vector of the II-nd kind f. Therefore f^* with components $f_{14}, f_{24}, f_{34}, f_{23}, f_{31}, f_{21}$ will be called the *dual spacetime vector* of f.

6^0. If w and s are two spacetime vectors of the I-st kind then by $w\overline{s}$ (as well as by $s\overline{w}$) we will be understand the combination of their components

$$w_1 s_1 + w_2 s_2 + w_3 s_3 + w_4 s_4. \tag{43}$$

Under a Lorentz transformation A this combination is invariant because of $(wA)(\overline{As}) = w\overline{s}$. When $w\overline{s} = 0$ w and s are *normal* to each other.

Two spacetime vectors of the I-st kind w, s form a 2×4 matrix

$$\begin{vmatrix} w_1, & w_2, & w_3, & w_4 \\ s_1, & s_2, & s_3, & s_4 \end{vmatrix}.$$

It can be seen immediately that the system of six quantities

$$\begin{array}{cccc} w_2 s_3 - w_3 s_2, & w_3 s_1 - w_1 s_3, & w_1 s_2 - w_2 s_1, & w_1 s_4 - w_4 s_1, \\ & w_2 s_4 - w_4 s_2, & w_3 s_4 - w_4 s_3 \end{array} \tag{44}$$

behaves under the Lorentz transformations as a spacetime vector of the II-nd type. This vector of the II-nd kind with the components (44) is denoted by $[w, s]$. It is easily seen that $\mathrm{Det}^{\frac{1}{2}}[w, s] = 0$. The dual vector of $[w, s]$ will be written as $[w, s]^{*}$.

If w is a spacetime vector of the I-st kind, and f is a spacetime vector of the II-st kind, then the product wf is a 1×4 matrix. Under a Lorentz transformation A the quantity w goes into $w' = w\mathsf{A}$, f into $f' = \mathsf{A}^{-1}f\mathsf{A}$; then $w'f' = w\mathsf{A}\mathsf{A}^{-1}f\mathsf{A} = (wf)\mathsf{A}$, i.e. wf transforms as a spacetime vector of the I-st kind.

If w is a spacetime vector of the I-st kind and f is a vector of the II-nd kind, the important identity can be easily verified

$$[w, wf] + [w, wf^{*}]^{*} = (w\overline{w})f. \tag{45}$$

The sum of the two space time vectors of the II-nd kind on the left side should be understood in the sense of the addition of two alternating matrices.

Concretely for $w_1 = 0, w_2 = 0, w_3 = 0, w_4 = i$ we have

$$wf = |if_{41}, if_{42}, if_{43}, 0|; \qquad wf^{*} = |if_{32}, if_{13}, if_{21}, 0|;$$
$$[w, wf] = 0, 0, 0, f_{41}, f_{42}, f_{43}; \qquad [w, wf^{*}] = 0, 0, 0, f_{32}, f_{13}, f_{21},$$

and in this particular case the relation(45) is satisfied, which is sufficient to show that the identity (45) holds generally because this relation has a covariant character under a Lorentz transformation, and is homogeneous in $w_1 = 0, w_2 = 0, w_3 = 0, w_4 = i$.

After these preparations, we deal first with the equations (C,) (D,) (E) by which the constants ε, μ, σ are introduced.

Instead of the space vector $\mathfrak{w}$, the velocity of matter, we will introduce, as in §8, the spacetime vector of the I-st kind w with its four components

$$w_1 = \frac{\mathfrak{w}_x}{\sqrt{1 - \mathfrak{w}^2}}, \quad w_2 = \frac{\mathfrak{w}_y}{\sqrt{1 - \mathfrak{w}^2}}, \quad w_3 = \frac{\mathfrak{w}_z}{\sqrt{1 - \mathfrak{w}^2}}, \quad w_4 = \frac{i}{\sqrt{1 - \mathfrak{w}^2}},$$

where

$$(w\overline{w}) = w_1^2 + w_2^2 + w_3^2 + w_4^2 = -1 \tag{46}$$

and $-iw_4 > 0$.

Now F and f will be again understood as the occurring in the fundamental equations spacetime vectors of the II-nd kind, $\mathfrak{M}, -i\mathfrak{E}$, and $\mathfrak{m}, -i\mathfrak{e}$.

Through $\Phi = -wF$ we have another spacetime vector of the I-st kind with components

$$
\begin{aligned}
\Phi_1 &= & w_2 F_{12} + w_3 F_{13} + w_4 F_{14} \\
\Phi_2 &= w_1 F_{21} & + w_3 F_{23} + w_4 F_{24} \\
\Phi_3 &= w_1 F_{31} + w_2 F_{32} & + w_4 F_{34} \\
\Phi_4 &= w_1 F_{41} + w_2 F_{24} + w_3 F_{34} &
\end{aligned}
$$

The first three quantities Φ_1, Φ_2, Φ_3 are the x-,y-,z-components of the space-vector

$$
\frac{\mathfrak{E} + [\mathfrak{w}\mathfrak{M}]}{\sqrt{1 - \mathfrak{w}^2}}, \tag{47}
$$

and

$$
\Phi_4 = \frac{i(\mathfrak{w}\mathfrak{E})}{\sqrt{1 - \mathfrak{w}^2}}. \tag{48}
$$

Since the F is an alternating matrix, we obviously have

$$
w\overline{\Phi} = w_1 \Phi_1 + w_2 \Phi_2 + w_3 \Phi_3 + w_4 \Phi_4 = 0, \tag{49}
$$

which shows that the vector Φ is normal to w; this relation can be also written as

$$
\Phi_4 = i(\mathfrak{w}_x \Phi_1 + \mathfrak{w}_y \Phi_2 + \mathfrak{w}_z \Phi_3). \tag{50}
$$

I will call the spacetime vector of the I-st kind Φ the *electric rest force*.

Analogous relations between such as between $-w, F, \mathfrak{E}, \mathfrak{M}, \mathfrak{w}$ hold also between $-wf, \mathfrak{e}, \mathfrak{m}, \mathfrak{w}$, and, in particular, $-wf$ is normal to w. The relation (C) can be now replaced with

$$
wf = \varepsilon wF. \tag{C}
$$

a formula which, although provides four equations for the relevant components, but such that the fourth, due to (50), is a consequence of the first three.

Let us now form the time-space vector I-st kind $\Psi = iwf^*$ whose components are

$$
\begin{aligned}
\Psi_1 &= -i(\qquad\quad w_2 f_{34} + w_3 f_{42} + w_4 f_{23}) \\
\Psi_2 &= -i(w_1 f_{43} \qquad\quad + w_3 f_{14} + w_4 f_{31}) \\
\Psi_3 &= -i(w_1 f_{24} + w_2 f_{41} \qquad\quad + w_4 f_{12}) \\
\Psi_4 &= -i(w_1 f_{32} + w_2 f_{13} + w_3 f_{21} \qquad\quad).
\end{aligned}
$$

The first three Ψ_1, Ψ_2, Ψ_3, are the x-, y-, z-components of the space-vector

$$
\frac{\mathfrak{m} - [\mathfrak{we}]}{\sqrt{1 - \mathfrak{w}^2}}, \tag{51}
$$

and also

$$
\Psi_4 = \frac{i(\mathfrak{wm})}{\sqrt{1 - \mathfrak{w}^2}}. \tag{52}
$$

and between them there exists the relation

$$
w\overline{\Psi} = w_1\Psi_1 + w_2\Psi_2 + w_3\Psi_3 + w_4\Psi_4 = 0, \tag{53}
$$

which can be also written as

$$
\Psi_4 = i(\mathfrak{w}_x\Psi_1 + \mathfrak{w}_y\Psi_2 + \mathfrak{w}_z\Psi_3). \tag{54}
$$

which shows that the vector Ψ is normal to w. I will call the spacetime vector of the I-st kind Ψ the *magnetic rest force*.

Relations analogous to the ones between $imf^*, \mathfrak{m}, \mathfrak{e}, \mathfrak{w}$ hold also between $imF^*, \mathfrak{M}, \mathfrak{E}, \mathfrak{w}$, and the relation (D) can be replaced with

$$
wF^* = \mu w f^*. \tag{D}
$$

The relations C and D can be used to express the field vectors F and f in terms of Φ and Ψ. We have

$$
wF = -\Phi, \qquad wF^* = -i\mu\Psi, \qquad wf = -\varepsilon\Phi, \qquad wf^* = -i\Psi,
$$

and applying the rule (45) by taking into account (46) leads to

$$F = [w, \Phi] + i\mu[w, \Psi]^*, \tag{55}$$
$$f = \varepsilon[w, \Phi] + i[w, \ \Psi]^*, \tag{56}$$

i.e. explicitly

$$F_{12} = (w_1\Phi_2 - w_2\Phi_1) + i\mu(w_3\Psi_4 - w_4\Psi_3), \text{ etc.}$$
$$f_{12} = \varepsilon(w_1\Phi_2 - w_2\Phi_1) + i(w_3\Psi_4 - w_4\Psi_3), \text{ etc.}$$

We also find the spacetime vector of the II-nd kind $[\Phi, \Psi]$ with the six components

$$\Phi_2\Psi_3 - \Phi_3\Psi_2, \qquad \Phi_3\Psi_1 - \Phi_1\Psi_3, \qquad \Phi_1\Psi_2 - \Phi_2\Psi_1,$$
$$\Phi_1\Psi_4 - \Phi_4\Psi_1, \qquad \Phi_2\Psi_4 - \Phi_4\Psi_2, \qquad \Phi_3\Psi_4 - \Phi_4\Psi_3.$$

Then the corresponding spacetime vector of the I-st kind

$$w[\Phi, \Psi] = -(w\overline{\Psi})\Phi + (w\overline{\Phi})\Psi$$

vanishes identically due to (49) and (53). Let us now consider the spacetime vector of the I-st kind

$$\Omega = iw[\Phi, \Psi]^* \tag{57}$$

with the components

$$\Omega_1 = -i \begin{vmatrix} w_2, & w_2, & w_3 \\ \Phi_2, & \Phi_3, & \Phi_4 \\ \Psi_2, & \Psi_3, & \Psi_4 \end{vmatrix}, \quad \text{etc.}$$

and by applying rule (45) it follows

$$[\Phi, \Psi] = i[w, \Omega]^* \tag{58}$$

i.e. explicitly

$$\Phi_1\Psi_2 - \Phi_2\Psi_1 = i(w_3\Omega_4 - w_4\Omega_3), \quad \text{etc.}$$

The vector Ω obviously satisfies the relation

$$(w, \overline{\Omega}) = w_1\Omega_1 + w_2\Omega_2 + w_3\Omega_3 + w_4\Omega_4 = 0, \tag{59}$$

which can be written as

$$\Omega_4 = i(\mathfrak{w}_x\Omega_1 + \mathfrak{w}_y\Omega_2 + \mathfrak{w}_z\Omega_3),$$

which means that Ω is *normal to* w. If $\mathfrak{w} = 0$ one has $\Phi_4 = 0$, $\Psi_4 = 0$, $\Omega_4 = 0$
and

$$\Omega_1 = \Phi_2\Psi_3 - \Phi_3\Psi_2, \quad \Omega_2 = \Phi_3\Psi_1 - \Phi_1\Psi_3, \quad \Omega_3 = \Phi_1\Psi_2 - \Phi_2\Psi_1. \tag{60}$$

I will call the spacetime vector of the I-st kind *rest ray*.
    Regarding the relation (E), which introduces the conductivity σ, we see
first that

$$-w\overline{s} = -(w_1 s_1 + w_2 s_2 + w_3 s_3 + w_4 s_4) = \frac{-\{\mathfrak{w}\}\mathfrak{s}_\mathfrak{w} + \varrho}{\sqrt{1 - \mathfrak{w}^2}} = \varrho'$$

is the *rest density* of electricity (see §8 and the end of §4). Then I will call
the spacetime vector of the I-st kind

$$s + (w\overline{s})w, \tag{61}$$

which due to $w\overline{w} = -1$ is obviously *normal to* w, the *rest current*. Let us
regard the first three components of this vector as the x-,y-,z-components
of a space vector, so the component in the direction of $\mathfrak{w}$ is:

$$\mathfrak{s}_\mathfrak{w} - \frac{|\mathfrak{w}|\varrho'}{\sqrt{1 - \mathfrak{w}^2}} = \frac{\mathfrak{s}_\mathfrak{w} - |\mathfrak{w}|\varrho}{1 - \mathfrak{w}^2} = \frac{\mathfrak{J}_\mathfrak{w}}{1 - \mathfrak{w}^2},$$

and the component in any perpendicular to $\mathfrak{w}$ direction $\overline{\mathfrak{w}}$

$$\mathfrak{s}_{\overline{\mathfrak{w}}} = \mathfrak{J}_{\overline{\mathfrak{w}}};$$

this space vector is analogous to the space vector $\mathfrak{J} = \mathfrak{s} - \varrho\mathfrak{w}$ which we called
in §8 the conduction current.
    Now, by comparison with $\Phi = -wF$ the relation (E) can be written in
the form:

$$s + (w\bar{s})w = -\sigma w F. \qquad \{E\}$$

This formula contains four equations, but since on both sides there are spacetime vectors of the I-st kind which are normal to w, the fourth is a consequence of the first three.

Finally, we will transform the differential equations (A) and (B) into a typical form.

§12.

The differential operator Lor

A 4×4 matrix

$$S = \begin{vmatrix} S_{11}, & S_{12}, & S_{13}, & S_{14} \\ S_{21}, & S_{22}, & S_{23}, & S_{24} \\ S_{31}, & S_{32}, & S_{33}, & S_{34} \\ S_{41}, & S_{42}, & S_{43}, & S_{44} \end{vmatrix} = |S_{hk}|, \qquad (62)$$

which obeys the rule that under a Lorentz transformation A it is always replaced by $\overline{A}SA$, can be called a *spacetime matrix* of the II-nd kind. Such a matrix is one in particular:

the alternating matrix f, which corresponds to the spacetime vector of the II-nd kind f,

the product fF of two such matrices f and F which under a transformation A is replaced by $(A^{-1}fA)(A^{-1}FA) = A^{-1}fFA$,

the 4×4 matrix with the elements $S_{hk} = w_h \Omega_k$ which are the products of the components w_1, w_2, w_3, w_4 and $\Omega_1, \Omega_2, \Omega_3, \Omega_4$ of two spacetime vectors of the I-st kind,

finally, a 4×4 matrix proportional to the unit matrix, i.e. a multiple L of the unit matrix, whose elements in the main diagonal have the same value L and the remaining elements are all zero.

Here we always deal with functions of spacetime points x, y, c, it and it can be advantageous to use a 1×4 *matrix formed from the differential symbols*

$$\begin{vmatrix} \dfrac{\partial}{\partial x}, & \dfrac{\partial}{\partial y}, & \dfrac{\partial}{\partial z}, & \dfrac{\partial}{i\partial t} \end{vmatrix},$$

or also

$$\left| \frac{\partial}{\partial x_1}, \frac{\partial}{\partial x_2}, \frac{\partial}{\partial x_3}, \frac{\partial}{\partial x_4} \right|. \tag{63}$$

For this matrix, I will need the *abbreviation* lor.

If S as in (62) is a spacetime matrix of the II-nd kind, then by the rule for product formation of matrices, under lor S we will understand the 1×4 matrix

$$|K_1,\ K_2,\ K_3,\ K_4|,$$

with the expressions

$$K_k = \frac{\partial S_{1k}}{\partial x_1} + \frac{\partial S_{2k}}{\partial x_2} + \frac{\partial S_{3k}}{\partial x_3} + \frac{\partial S_{4k}}{\partial x_4} \qquad (k = 1, 2, 3, 4). \tag{64}$$

When a Lorentz transformation A introduces a new reference system x'_1, x'_2, x'_3, x'_4 for the spacetime point, so the analogous operator can be constructed

$$\mathrm{lor}' = \left| \frac{\partial}{\partial x'_1}, \frac{\partial}{\partial x'_2}, \frac{\partial}{\partial x'_3}, \frac{\partial}{\partial x'_4} \right|.$$

As S transforms into $S' = \overline{\mathsf{A}}S\mathsf{A} = |S'_{hk}|$, so $\mathrm{lor}'S'$ should be a 1×4 matrix, with the expressions

$$K'_k = \frac{\partial S'_{1k}}{\partial x'_1} + \frac{\partial S'_{2k}}{\partial x'_2} + \frac{\partial S'_{3k}}{\partial x'_3} + \frac{\partial S'_{4k}}{\partial x'_4} \qquad (k = 1, 2, 3, 4).$$

Now for the differentiation of any function of a spacetime point applies the rule

$$\frac{\partial}{\partial x'_k} = \frac{\partial}{\partial x_1}\frac{\partial x_1}{\partial x'_k} + \frac{\partial}{\partial x_2}\frac{\partial x_2}{\partial x'_k} + \frac{\partial}{\partial x_3}\frac{\partial x_3}{\partial x'_k} + \frac{\partial}{\partial x_4}\frac{\partial x_4}{\partial x'_k} =$$

$$= \frac{\partial}{\partial x_1}\alpha_{1k} + \frac{\partial}{\partial x_2}\alpha_{2k} + \frac{\partial}{\partial x_3}\alpha_{3k} + \frac{\partial}{\partial x_4}\alpha_{4k},$$

which in an easily understandable way symbolically is

$$\mathrm{lor}' = \mathrm{lor}\ \mathsf{A}$$

from which it follows immediately

$$\text{lor}'\, S' = \text{lor}\,(A(A^{-1}SA)) = (\text{lor}\, S)A, \tag{65}$$

i.e. if S is a spacetime matrix of the II-nd kind, then lor S *transforms as a spacetime vector of the I-st kind.*

If L is a multiple of the unit matrix, then lor L is *the matrix with elements*

$$\left| \frac{\partial L}{\partial x_1},\ \frac{\partial L}{\partial x_2},\ \frac{\partial L}{\partial x_3},\ \frac{\partial L}{\partial x_4} \right|. \tag{66}$$

If $s = |s_1, s_2, s_3, s_4|$ is a spacetime vector of the I-st kind, then

$$\text{lor}\,\overline{s} = \frac{\partial s_1}{\partial x_1} + \frac{\partial s_2}{\partial x_2} + \frac{\partial s_3}{\partial x_3} + \frac{\partial s_4}{\partial x_4}. \tag{67}$$

If upon application of a Lorentz transformation A the signs lor$'$, and s' are used in place of lor and s, it follows that

$$\text{lor}'\,\overline{s}' = (\text{lor}\, A)(\overline{A\overline{s}}) = \text{lor}\,\overline{s},$$

i.e. lor $\overline{s}$ *is an invariant under the Lorentz transformations.*

In all these relations the operator lor *plays the role of a spacetime vector of the I-st kind*

If f is a spacetime vector of the II-nd kind, lor f is a spacetime vector of the I-st kind with components

$$\frac{\partial f_{12}}{\partial x_2} + \frac{\partial f_{13}}{\partial x_3} + \frac{\partial f_{14}}{\partial x_4},$$

$$\frac{\partial f_{21}}{\partial x_1} \qquad\quad + \frac{\partial f_{23}}{\partial x_3} + \frac{\partial f_{24}}{\partial x_4},$$

$$\frac{\partial f_{31}}{\partial x_1} + \frac{\partial f_{32}}{\partial x_2} \qquad\quad + \frac{\partial f_{34}}{\partial x_4},$$

$$\frac{\partial f_{41}}{\partial x_1} + \frac{\partial f_{42}}{\partial x_2} + \frac{\partial f_{43}}{\partial x_3} \qquad\quad .$$

After that the system of differential equations (A) can be expressed in the short form

$$\text{lor}\, f = -s. \tag{A}$$

In exactly the same way the system of differential equations (B) can be written as:

$$\text{lor } F^* = 0. \tag{B}$$

In regard to definition (67) the compounds formed from lor $(\overline{\text{lor } f})$ and lor $(\overline{\text{lor } F^*})$ vanish identically when f and F^* are alternating matrices. After this, the relationship between the components of the current s

$$\frac{\partial s_1}{\partial x_1} + \frac{\partial s_2}{\partial x_2} + \frac{\partial s_3}{\partial x_3} + \frac{\partial s_4}{\partial x_4} = 0 \tag{68}$$

follows from {A} while the relation

$$\text{lor } (\overline{\text{lor } F^*}) = 0 \tag{69}$$

has the meaning that *out of the four equations for the evolution of the field vectors in* {B} *only three are independent.*

I now summarize the results together:

Let w denote the spacetime vector of the I-st kind $\frac{\mathfrak{w}}{\sqrt{1-\mathfrak{w}^2}}, \frac{i}{\sqrt{1-\mathfrak{w}^2}}$ ($\mathfrak{w}$ is the velocity of matter), F – *a spacetime vector of the II-nd kind $\mathfrak{M}, -\mathfrak{E}$ ($\mathfrak{M}$* – magnetic induction, $\mathfrak{E}$ – electric force), f – *a spacetime vector of the II-nd kind $\mathfrak{m}, -\mathfrak{e}$ ($\mathfrak{m}$* – magnetic force, $\mathfrak{e}$ – electric induction), s – a spacetime vector of the I-st kind $\mathfrak{s}, i\varrho$ (ϱ – density of the electrical charge, $\mathfrak{s} - \varrho\mathfrak{w}$ – conductivity current), ε – *dielectric constant,* μ – *magnetic permeability,* σ – *conductivity, then these are* (with the explained symbols of matrices in §10 and §11) *the fundamental equations for the electromagnetic processes in moving bodies*

$$\text{lor } f = -s \tag{A}$$

$$\text{lor } F^* = 0 \tag{B}$$

$$wf = \varepsilon wF \tag{C}$$

$$wF^* = \mu wf^* \tag{D}$$

$$s + (w\overline{s})w = -\sigma wF. \tag{E}$$

Here $w\overline{w} = -1$ and the spacetime vectors of the I-st kind wF, wf, wF^*, wf^*, $s + (w\overline{s})s$ are normal to w and finally for the system of equations for $\{B\}$ the relationship

$$\text{lor}\,(\overline{\text{lor}\,F^*}) = 0$$

holds.

In view of the the last relation, we see that we have the necessary independent equations in order to determine the processes in appropriate boundary conditions, *if the motion of matter, expressed by the vector $\mathfrak{w}$ as a function of x, y, z, t is given.*

§13.

The product of the field vectors f F

Finally, we ask for the laws determining the vector w as a function of x, y, z, t. In the investigations relating to that, of interest are expressions resulting from the multiplication of two alternating matrices

$$f = \begin{vmatrix} 0, & f_{12}, & f_{13}, & f_{14} \\ f_{21}, & 0, & f_{23}, & f_{24} \\ f_{31}, & f_{32}, & 0, & f_{34} \\ f_{41}, & f_{42}, & f_{43}, & 0 \end{vmatrix}, \quad F = \begin{vmatrix} 0, & F_{12}, & F_{13}, & F_{14} \\ F_{21}, & 0, & F_{23}, & F_{24} \\ F_{31}, & F_{32}, & 0, & F_{34} \\ F_{41}, & F_{42}, & F_{43}, & 0 \end{vmatrix}.$$

The matrix product $f\,F$ can be represented in the form

$$fF = \begin{vmatrix} S_{11} - L, & S_{12}, & S_{13}, & S_{14} \\ S_{21}, & S_{22} - L, & S_{23}, & S_{24} \\ S_{31}, & S_{32}, & S_{33} - L, & S_{34} \\ S_{41}, & S_{42}, & S_{43}, & S_{44} - L \end{vmatrix}, \tag{70}$$

such that

$$S_{11} + S_{22} + S_{33} + + S_{44} = 0. \tag{71}$$

L is the symmetrical combination of all indices $1, 2, 3, 4$

$$L = \frac{1}{2}(f_{23}F_{23} + f_{31}F_{31} + f_{12}F_{12} + f_{14}F_{14} + f_{24}F_{24} + f_{34}F_{34}), \tag{72}$$

and then

$$S_{11} = \frac{1}{2}(f_{23}F_{23} + f_{34}F_{34} + f_{42}F_{42} - f_{12}F_{12} - f_{13}F_{13} - f_{14}F_{14}), \tag{73}$$

$$S_{12} = f_{13}F_{32} + f_{14}F_{42}, \quad \text{etc.}$$

As I express the conditions of reality, I want to write

$$S = \begin{vmatrix} S_{11}, & S_{12}, & S_{13}, & S_{14} \\ S_{21}, & S_{22}, & S_{23}, & S_{24} \\ S_{31}, & S_{32}, & S_{33}, & S_{34} \\ S_{41}, & S_{42}, & S_{43}, & S_{44} \end{vmatrix} = \begin{vmatrix} X_x, & Y_x, & Z_x, & -iT_x \\ X_y, & Y_y, & Z_y, & -iT_y \\ X_z, & Y_z, & Z_z, & -iT_z \\ -iX_t, & -iY_t, & -iZ_t, & T_t \end{vmatrix}, \tag{74}$$

where

$$X_x = \frac{1}{2}(\mathfrak{m}_x\mathfrak{M}_x - \mathfrak{m}_y\mathfrak{M}_y - \mathfrak{m}_z\mathfrak{M}_z + \mathfrak{e}_x\mathfrak{E}_x - \mathfrak{e}_y\mathfrak{E}_y - \mathfrak{e}_z\mathfrak{E}_z)$$

$$X_y = \mathfrak{m}_x\mathfrak{M}_y + \mathfrak{e}_y\mathfrak{E}_x, \qquad Y_x = \mathfrak{m}_y\mathfrak{M}_x + \mathfrak{e}_x\mathfrak{E}_y, \qquad \text{etc.}$$

$$X_t = \mathfrak{e}_y\mathfrak{M}_z - \mathfrak{e}_z\mathfrak{M}_y, \tag{75}$$

$$T_y = \mathfrak{m}_z\mathfrak{E}_y - \mathfrak{m}_y\mathfrak{E}_z, \qquad \text{etc.}$$

$$T_t = \frac{1}{2}(\mathfrak{m}_x\mathfrak{M}_x + \mathfrak{m}_y\mathfrak{M}_y + \mathfrak{m}_z\mathfrak{M}_z + \mathfrak{e}_x\mathfrak{E}_x + \mathfrak{e}_y\mathfrak{E}_y + \mathfrak{e}_z\mathfrak{E}_z)$$

and also

$$L = \frac{1}{2}(\mathfrak{m}_x\mathfrak{M}_x + \mathfrak{m}_y\mathfrak{M}_y + \mathfrak{m}_z\mathfrak{M}_z - \mathfrak{e}_x\mathfrak{E}_x - \mathfrak{e}_y\mathfrak{E}_y - \mathfrak{e}_z\mathfrak{E}_z) \tag{76}$$

are all real. In the theories of bodies at rest the combinations $X_x, X_y, X_z,$ $Y_x, Y_y, Y_z, Z_x, Z_y, Z_z$ are known as "Maxwell's Stresses", the quantities $T_x,$ T_y, T_z are known as the "Poynting's vector," T_t as the "electromagnetic energy density," and L as the "Lagrangian function."

On the other hand, we can take the dual matrices of f and F and multiply them in reverse order

$$F^*f^* = \begin{vmatrix} -S_{11} - L, & -S_{12}, & -S_{13}, & -S_{14} \\ -S_{21}, & -S_{22} - L, & -S_{23}, & -S_{24} \\ -S_{31}, & -S_{32}, & -S_{33} - L, & -S_{34} \\ -S_{41}, & -S_{42}, & -S_{43}, & -S_{44} - L \end{vmatrix}, \tag{77}$$

and we can hereafter set

$$fF = S - L, \qquad F^*f^* = -S - L \qquad (78)$$

where by L, we mean the matrix $L \cdot 1$ that is proportional to the unit matrix, i.e. the matrix with elements

$$|Le_{hk}| \qquad \begin{pmatrix} e_{hh} = 1, & e_{hk} = 0, & h \neq k \\ & h, k = 1, 2, 3, 4 \end{pmatrix}.$$

As $SL = LS$ we conclude further

$$F^*f^*fF = (-S - L)(S - L) = -SS + L^2,$$

from here, since $f^*f = \mathrm{Det}^{\frac{1}{2}} f$, $F^*F = \mathrm{Det}^{\frac{1}{2}} F$, we find the interesting relation

$$SS = L^2 - \mathrm{Det}^{\frac{1}{2}} f \, \mathrm{Det}^{\frac{1}{2}} F, \qquad (79)$$

i.e. *the product of the matrix S with itself is proportional to the unit matrix.* relations

$$S_{h1}S_{1k} + S_{h2}S_{2k} + S_{h3}S_{3k} + S_{h4}S_{4k} = 0 \qquad (80)$$

for unequal indices h, k from the series $1, 2, 3, 4$ and

$$S_{h1}S_{1h} + S_{h2}S_{2h} + S_{h3}S_{3h} + S_{h4}S_{4h} = L^2 - \mathrm{Det}^{\frac{1}{2}} f \, \mathrm{Det}^{\frac{1}{2}} F \qquad (81)$$

for $h = 1, 2, 3, 4$.

Now if the quantities F and f in (72), (73) are expressed, by taking into account (55), (56) and (57), through the *electrical rest force* Φ, *the magnetic rest force* Ψ *and the rest ray* Ω, we arrive at the expressions:

$$L = -\frac{1}{2}\varepsilon \Phi \overline{\Phi} + \frac{1}{2}\mu \Psi \overline{\Psi}, \qquad (82)$$

$$S_{hk} = -\frac{1}{2}\varepsilon\Phi\overline{\Phi}e_{hk} + \frac{1}{2}\mu\Psi\overline{\Psi}e_{hk}$$
$$+ \varepsilon(\Phi_h\Phi_k - \Phi\overline{\Phi}w_hw_k) + \mu(\Psi_h\Psi_k - \Psi\overline{\Psi}w_hw_k) \quad (83)$$
$$- \Omega_h w_k - \varepsilon\mu w_h\Omega_k \quad (h, k = 1, 2, 3, 4),$$

where

$$\Phi\overline{\Phi} = \Phi_1^2 + \Phi_2^2 + \Phi_3^2 + \Phi_4^2, \qquad \Psi\overline{\Psi} = \Psi_1^2 + \Psi_2^2 + \Psi_3^2 + \Psi_4^2,$$

$$e_{hh} = 1, \qquad e_{hk} = 0 \quad (h \neq k).$$

Namely, in any case, the right side of (82) as well as L is an invariant under Lorentz transformations and the right side of (83) as well as S_{hk} represent the elements of a 4×4 spacetime matrix of the II-nd kind. Considering this, it is enough for the relations (82) and (83) to hold generally, if they are verified only for the case $w_1 = 0$, $w_2 = 0$, $w_3 = 0$, $w_4 = i$. But in this case $\mathfrak{w} = 0$ equations (82), (83) follow, on the one hand, from (47), (51), (60), and, on the other hand, from $\mathfrak{e} = \varepsilon\mathfrak{E}$, $\mathfrak{M} = \mu\mathfrak{m}$ directly due to the equations (75), (76).

The expression on the right in (81)

$$= \left(\frac{1}{2}(\mathfrak{m}\mathfrak{M} - \mathfrak{e}\mathfrak{E})\right)^2 + (\mathfrak{e}\mathfrak{m})(\mathfrak{E}\mathfrak{M}),$$

due to the relations $(\mathfrak{e}\mathfrak{m}) = \varepsilon\Phi\overline{\Psi}$, $(\mathfrak{E}\mathfrak{M}) = \mu\Phi\overline{\Psi}$ is always ≥ 0; the square root of it, which is ≥ 0, will be in accordance with (79) denoted by $\mathrm{Det}^{\frac{1}{4}}S$.

For $\overline{S}$, *the transposed matrix of* S, as $\overline{f} = -f$ and $\overline{F} = -F$, it follows from (78) that

$$Ff = \overline{S} - L, \qquad F^*f^* = -\overline{S} - L \quad (84)$$

Then

$$S - \overline{S} = |S_{hk} - S_{kh}|$$

is an alternating matrix and also means a spacetime vector of the II-nd kind. From the expressions (83) we immediately have

$$S - \overline{S} = -(\varepsilon\mu - 1)[w, \Omega], \quad (85)$$

from where it is also deduced see (57),(58))

$$w(S - \overline{S})^* = 0, \tag{86}$$

$$w(S - \overline{S}) = (\varepsilon\mu - 1)\Omega. \tag{87}$$

When at a spacetime point the matter is at rest, i.e. $\mathfrak{w} = 0$, *then (86) reduced to the following equations*

$$Z_y = Y_z, \qquad X_z = Z_x, \qquad Y_x = X_y,$$

and then one has from (83):

$$T_x = \Omega_1, \qquad T_y = \Omega_2, \qquad T_z = \Omega_3,$$
$$X_t = \varepsilon\mu\Omega_1, \qquad Y_t = \varepsilon\mu\Omega_2, \qquad Z_t = \varepsilon\mu\Omega_3$$

Now by a suitable rotation of the spatial coordinate system x, y, z about the origin, we can have

$$Z_y = Y_z = 0, \qquad X_z = Z_x = 0, \qquad Y_x = X_y = 0.$$

According to (71) one has

$$X_x + Y_y + Z_z + T_t = 0, \tag{88}$$

and always $T > 0$ as seen from (83). In the special case when Ω also vanishes, one obtains from (81)

$$X_x^2 = Y_y^2 = Z_z^2 = T_t^2 = (\mathrm{Det}^{\frac{1}{4}} S)^2$$

and if T_t and one of the three magnitudes X_x, Y_y, Z_z are $= +\mathrm{Det}^{\frac{1}{4}} S$, the remaining two are $= -\mathrm{Det}^{\frac{1}{4}} S$. If Ω does not vanish, we can take, for example, $\Omega_3 \neq 0$ and then according to (80) in particular

$$T_z X_t = 0, \qquad T_z Y_t = 0, \qquad Z_z T_z + T_z T_t = 0,$$

and therefore $\Omega_1 = 0, \Omega_2 = 0, Z_z = -T_t$. From (81) and in view of (88) then it follows

$$X_x = -Y_y = \pm \mathrm{Det}^{\frac{1}{4}} S,$$

$$-Z_z = T_t = \sqrt{\operatorname{Det}^{\frac{1}{2}} S + \varepsilon \mu \Omega_3^2} > \operatorname{Det}^{\frac{1}{4}} S.$$

Finally, of particular importance is the spacetime vector of the I-st kind

$$K = \operatorname{lor} S, \tag{89}$$

for which we now want to prove an important transformation.

According to (78) $S = L + fF$ and it follows that

$$\operatorname{lor} S = \operatorname{lor} L + \operatorname{lor} fF.$$

The symbol lor represents differentiation, which in lor fF applies, on the one hand, to the components of f, and, on the other hand, to the components of F. Accordingly, lor fF is additively decomposed into a first and a second part. The first part is obviously the matrix product (lor f) F, where lor f is interpreted by itself as a 4×4 matrix. The second term is that part from lor fF, in which the differentiations affect only the components of F. Now we see from (78)

$$fF = -F^* f^* - 2L,$$

consequently the second part of lor fF is $-(\operatorname{lor} F)^* f* +$ the part of $-2 \operatorname{lor} L$, in which the differentiations apply only to the components of F. After this, the result is

$$\operatorname{lor} S = (\operatorname{lor} f)F - (\operatorname{lor} F^*)f^* + N, \tag{90}$$

where N is a vector with components

$$N_h = \frac{1}{2} \left(\frac{\partial f_{23}}{\partial x_h} F_{23} + \frac{\partial f_{31}}{\partial x_h} F_{31} + \frac{\partial f_{12}}{\partial x_h} F_{12} + \frac{\partial f_{14}}{\partial x_h} F_{14} + \frac{\partial f_{24}}{\partial x_h} F_{24} + \frac{\partial f_{34}}{\partial x_h} F_{34} \right)$$
$$- \frac{1}{2} \left(f_{23} \frac{\partial F_{23}}{\partial x_h} + f_{31} \frac{\partial F_{31}}{\partial x_h} + f_{12} \frac{\partial F_{12}}{\partial x_h} + f_{14} \frac{\partial F_{14}}{\partial x_h} + f_{24} \frac{\partial F_{24}}{\partial x_h} + f_{34} \frac{\partial F_{34}}{\partial x_h} \right)$$
$$(h = 1, 2, 3, 4).$$

Using the fundamental equations $\{A\}$ and $\{B\}$ transforms (90) into the *fundamental relation*

$$\operatorname{lor} S = -sF + N. \tag{91}$$

In the limiting case $\varepsilon = 1$, $\mu = 1$, when $f = F$, N vanishes identically.

In general, due to (55), (56) and with regard to the expression (82) for L and from (57) we arrive the following expressions of the components of N:

$$
\begin{aligned}
N_h = & -\frac{1}{2}\Phi\overline{\Phi}\frac{\partial\varepsilon}{\partial x_h} - \frac{1}{2}\Psi\overline{\Psi}\frac{\partial\mu}{\partial x_h} \\
& + (\varepsilon\mu - 1)\left(\Omega_1\frac{\partial w_1}{\partial x_h} + \Omega_2\frac{\partial w_2}{\partial x_h} + \Omega_3\frac{\partial w_3}{\partial x_h} + \Omega_4\frac{\partial w_4}{\partial x_h}\right)
\end{aligned}
\tag{92}
$$
$$
\text{for} \quad h = 1, 2, 3, 4.
$$

By using (59) and introducing the space vector $\mathfrak{W}$ with $\Omega_1, \Omega_2, \Omega_3$ as its x-, y-, z-components, then the third component of (92) can be concisely expressed as

$$
\frac{\varepsilon\mu - 1}{\sqrt{1 - \mathfrak{w}^2}}\left(\mathfrak{W}\frac{\partial\mathfrak{w}}{\partial x_h}\right),
\tag{93}
$$

where the bracket indicates the scalar product of two vectors therein.

§14.

The ponderomotive forces

In a more detailed form the equation $K = \mathrm{lor}\, S = -sF + N$ is represented by four equations:

$$
\begin{aligned}
K_1 = & \frac{\partial X_x}{\partial x} + \frac{\partial X_y}{\partial y} + \frac{\partial X_z}{\partial z} - \frac{\partial X_t}{\partial t} = \varrho\mathfrak{E}_x + \mathfrak{s}_y\mathfrak{M}_z - \mathfrak{s}_z\mathfrak{M}_y \\
& -\frac{1}{2}\Phi\overline{\Phi}\frac{\partial\varepsilon}{\partial x} - \frac{1}{2}\Psi\overline{\Psi}\frac{\partial\mu}{\partial x} + \frac{\varepsilon\mu - 1}{\sqrt{1 - \mathfrak{w}^2}}\left(\mathfrak{W}\frac{\partial\mathfrak{w}}{\partial x}\right),
\end{aligned}
\tag{94}
$$

$$
\begin{aligned}
K_2 = & \frac{\partial Y_x}{\partial x} + \frac{\partial Y_y}{\partial y} + \frac{\partial Y_z}{\partial z} - \frac{\partial Y_t}{\partial t} = \varrho\mathfrak{E}_y + \mathfrak{s}_z\mathfrak{M}_x - \mathfrak{s}_x\mathfrak{M}_z \\
& -\frac{1}{2}\Phi\overline{\Phi}\frac{\partial\varepsilon}{\partial y} - \frac{1}{2}\Psi\overline{\Psi}\frac{\partial\mu}{\partial y} + \frac{\varepsilon\mu - 1}{\sqrt{1 - \mathfrak{w}^2}}\left(\mathfrak{W}\frac{\partial\mathfrak{w}}{\partial y}\right),
\end{aligned}
\tag{95}
$$

$$K_3 = \frac{\partial Z_x}{\partial x} + \frac{\partial Z_y}{\partial y} + \frac{\partial Z_z}{\partial z} - \frac{\partial Z_t}{\partial t} = \varrho\mathfrak{E}_z + \mathfrak{s}_x\mathfrak{M}_y - \mathfrak{s}_y\mathfrak{M}_x$$
$$-\frac{1}{2}\Phi\overline{\Phi}\frac{\partial\varepsilon}{\partial z} - \frac{1}{2}\Psi\overline{\Psi}\frac{\partial\mu}{\partial z} + \frac{\varepsilon\mu - 1}{\sqrt{1 - \mathfrak{w}^2}}\left(\mathfrak{M}\frac{\partial\mathfrak{w}}{\partial z}\right),$$

(96)

$$\frac{1}{i}K_4 = -\frac{\partial T_x}{\partial x} - \frac{\partial T_y}{\partial y} - \frac{\partial T_z}{\partial z} - \frac{\partial T_t}{\partial t} = \mathfrak{s}_x\mathfrak{E}_x + \mathfrak{s}_y\mathfrak{E}_y + \sigma_z\mathfrak{E}_z$$
$$+\frac{1}{2}\Phi\overline{\Phi}\frac{\partial\varepsilon}{\partial t} + \frac{1}{2}\Psi\overline{\Psi}\frac{\partial\mu}{\partial t} - \frac{\varepsilon\mu - 1}{\sqrt{1 - \mathfrak{w}^2}}\left(\mathfrak{M}\frac{\partial\mathfrak{w}}{\partial t}\right).$$

(97)

It is now my opinion that in the electromagnetic processes the pondero-motive force which acts on the matter at a spacetime point x, y, z, t, calcu-lated for a unit volume, has as its x-,y-,z-components, the first three com-ponents of the spacetime vector

$$K + (w\overline{K})w,$$

(98)

which is normal to the spacetime vector w, and the energy law is represented by the fourth of the above equations.

Detailed reasons for this opinion will be reserved for a subsequent trea-tise; here I just want to give some supporting remarks on the mechanics of this opinion.

In the limiting case $\varepsilon = 1$, $\mu = 1$, $\sigma = 0$ we have $N = 0$, $\mathfrak{s} = \varrho\mathfrak{w}$, from where $w\overline{K} = 0$ and our approach coincides with those commonly used in the electron theory.

Appendix.

Mechanics and the relativity postulate.

Most likely it would be unsatisfactory if the new understanding of the concept of time, which is characterized by the freedom of the Lorentz trans-formations, is applied to only a branch of physics.

Now many authors say that classical mechanics is contrary to the relativity postulate, which is set here as the basis of electrodynamics.

In order to decide this let us concentrate on a special Lorentz transformation as determined by the equations (10), (11), (12) with a non-zero vector $\mathfrak{b}$ from any direction and a magnitude $q < 1$. For the moment we will not choose in advance any relation between the unit of length and the unit of time and will write not t, t', q, but $ct, ct', \frac{q}{c}$, where c is a positive constant and $q < c$. The equations mentioned above are therefore transformed into

$$\mathfrak{r}'_{\overline{\mathfrak{b}}} = \mathfrak{r}_{\overline{\mathfrak{b}}}, \qquad \mathfrak{r}'_{\mathfrak{b}} = \frac{c(\mathfrak{r}_{\mathfrak{b}} - qt)}{\sqrt{c^2 - q^2}}, \qquad t' = \frac{-q\mathfrak{r}_{\mathfrak{b}} + c^2 t}{c\sqrt{c^2 - q^2}},$$

where as we recall $\mathfrak{r}$ is the space vector with components x, y, x and $\mathfrak{r}'$ is the space vector with components x', y', z'.

If we keep $\mathfrak{v}$ constant, in the limiting case $c = \infty$ these equations become

$$\mathfrak{r}'_{\overline{\mathfrak{b}}} = \mathfrak{r}_{\overline{\mathfrak{b}}}, \qquad \mathfrak{r}'_{\mathfrak{b}} = \mathfrak{r}_{\mathfrak{b}} - qt, \qquad t' = t.$$

These new equations would now mean a transition from the current spatial coordinate system x, y, z to a different spatial coordinate systems x', y', z' with parallel axes, whose origin moves in a straight line with constant velocity with respect to the first, while the time parameter will remain completely unaffected.

Based on this observation one might say:

Classical mechanics postulates a covariance of physical laws for the group of homogeneous linear transformations of the expression

$$-x^2 - y^2 - z^2 + c^2 t^2, \tag{1}$$

with the setting $c = \infty$.

Now it would be absolutely confusing to find that in one branch of physics we find a covariance of the laws for the transformation of expression (1) with a finite value of c, whereas in another part – for $c = \infty$.

The situation that Newtonian mechanics deals with this covariance only for $c = \infty$ and not with the case of c as the speed of light, needs no explanation. Is it then possible to regard that traditional covariance for $c = \infty$ only as a first approximation, consistent with experience, of the exact covariance of natural laws holding for a certain finite value of c?

I want to say that by *reforming the mechanics such that instead of Newton's postulate of relativity with $c = \infty$ another one with a finite c holds,* even the axiomatic structure of mechanics appears to become considerably complete.

The ratio of the time unit to length unit is normalized in agreement with the relativity postulate with $c = 1$.

As I now want to introduce geometrical figures in the manifold of the four variables x, y, z, t it may be convenient not to take y, z into account, and to interpret x and t as any oblique coordinates in a plane.

A spacetime origin $O(x, y, z, t = 0, 0, 0, 0)$ is held unchanged under Lorentz transformations. The structure

$$-x^2 - y^2 - z^2 + t^2 = 1, \qquad t > 0, \tag{2}$$

is a *hyperboloidal shell*, which contains the spacetime point $A(x, y, z, t = 0, 0, 0, 1)$ and all points $A'(x', y', z', t')$ which are obtained from the first due to the Lorentz transformations.

The direction of a radius vector OA' from 0 to a point A' on the structure (2) and the directions of tangents at A' are normal to one another.

Let us trace a certain location of matter in its path at all times t. The entirety of the spacetime points x, y, z, t of the locations corresponding to different times t, I will call a *spacetime line*.

The task to determine the motion of matter should be understood in this way: *the direction of the spacetime line passing through each spacetime point should be found.*

A spacetime $P(x, y, z, t)$ can be *transformed to rest* by introducing through a Lorentz transformation a reference system x', y', z', t' such that its t'-axis should be chosen from the allowed directions OA' in such a way that it is parallel to the spacetime line passing through P. The space $t' = $const, which goes through P, is *normal* to the spacetime line through P. The increase dt of the time t at P corresponds to the increase

$$d\tau = \sqrt{dt^2 - dx^2 - dy^2 - dz^2} = dt\sqrt{1 - \mathfrak{w}^2} = \frac{dx_4}{w_4} \tag{3}$$

of the parameter t' introduced here[11]. The value of the integral

$$\int d\tau = \int \sqrt{-(dx_1^2 + dx_2^2 + dx_3^2 + dx_4^2)}$$

---

[11]The indices and the symbols $\mathfrak{w}$, w have the meaning adopted above (see §3 and §4).

taken along the spacetime line from any fixed initial point P^0 to a point P is called *proper time* (*Eigenzeit*), corresponding to the location of matter at the spacetime point P. (This is a generalization of the concept of *local time* (*Ortszeit*) used by Lorentz in the case of uniform motion.)

Consider a spatially extended body R^0 at a given time t^0, then the area consisting of the spacetime lines passing through the spacetime point R^0, t^0 is called a *spacetime thread* (*Raum-Zeitfaden*).

Let us have an analytical expression $\Theta(x, y, z, t)$ such that $\Theta(x, y, z, t) = 0$ is taken from each spacetime line of the thread in one point, with

$$-\left(\frac{\partial\Theta}{\partial x}\right)^2 - \left(\frac{\partial\Theta}{\partial y}\right)^2 - \left(\frac{\partial\Theta}{\partial z}\right)^2 + \left(\frac{\partial\Theta}{\partial t}\right)^2 > 0, \qquad \frac{\partial\Theta}{\partial t} > 0,$$

then the entirety Q of the points resulting from the intersection of the space of extended body R^0 and its spacetime thread will be called a *cross section* (*Querschnitt*) of the thread. At any point $P(x, y, z, t)$ of such across section, we can introduce through a Lorentz transformation a reference system x', y', z', t' in which

$$\frac{\partial\Theta}{\partial x'} = 0, \qquad \frac{\partial\Theta}{\partial y'} = 0, \qquad \frac{\partial\Theta}{\partial z'} = 0, \qquad \frac{\partial\Theta}{\partial t'} > 0.$$

The direction of such uniquely determined t'-axis will be called the *upper normal* of the cross section Q at the point P and the value $dJ = \iiint dx'dy'dz'$ for a neighborhood of P on the cross section will be called the *volume element* (*Inhaltselement*) of the cross section. In this sense, R^0, t^0 itself should be regarded as the cross section at $t = t^0$ which is normal to the t'-axis of the thread and the (ordinary) volume of the body R^0 should be called the *volume* (*Inhalt*) of this cross section.

By letting the space R^0 to converge toward a point we come to the concept of an *infinitely thin* spacetime thread. In such a thread a spacetime line will be regarded as the *main line* (*Hauptlinie*) and by *proper time of the thread* we will understand the proper time along this main line; by a *normal cross section* of the thread we will understand the space, which is normal to the main line at a given point on it.

We now formulate *the principle of conservation of mass*.

To every space R at a time t belongs a positive quantity, the mass in R at time t. When R converges to a point x, y, z, t the quotient of this mass and the volume of R approaches a limiting value $\mu(x, y, z, t)$, the *mass density* at the spacetime point x, y, z, t.

The principle of conservation of mass states that: *For an infinitely thin spacetime thread the product $\mu\,\mathrm{d}J$ of the mass density at a point x, y, z, t on the thread (i.e., on the main line of the thread) and the volume $\mathrm{d}J$ of the cross section, which is normal to the t-axis, is constant along the entire thread.*

Now, the volume $\mathrm{d}J_n$ of the normal cross section of the thread at point x, y, z, t is

$$\mathrm{d}J_n = \frac{1}{\sqrt{1 - \mathfrak{w}^2}}\mathrm{d}J = -iw_4\mathrm{d}J = \frac{\mathrm{d}t}{\mathrm{d}\tau}\mathrm{d}J \tag{4}$$

and it should be expected for

$$\nu = \frac{\mu}{-iw_4} = \mu\sqrt{1 - \mathfrak{w}^2} = \mu\frac{\mathrm{d}\tau}{\mathrm{d}t}, \tag{5}$$

to be defined as the *rest mass density* at the point x, y, z, t. Then the principle of conservation of mass can be also formulated as follows:

For an infinitely thin spacetime thread the product of the rest mass density and the volume of the normal cross section at a point on the thread is always constant along the entire thread.

In an arbitrary spacetime thread to a cross section Q^0 a second cross section Q^1 can be mounted which with Q^0 has points on the boundary of the thread, but with only this in common, and for the spacetime lines within the thread the values of t increase from Q^0 to Q^1. The finite area located between Q^0 and Q^1 will be called *spacetime sickle* (*Raum-Zeit-Sichel*), where Q^0 is the lower boundary and Q^1 – the upper boundary of the sickle.

Let us suppose that the thread is divided into many very thin spacetime threads, so every entry of a thin thread in the lower boundary of the sickle corresponds to an exit from the top, and each time for both cases the product $\nu\,\mathrm{d}J_n$, whose terms are defined by (4) and (5), has the same value.

Therefore the difference between the two integrals $\int \nu\,\mathrm{d}J_n$, the first extending over the upper, the second over the lower boundary of the sickle, vanishes. This difference can be found by using a known theorem of integral calculus and is equal to the integral

$$\iiiint \mathrm{lor}\,\nu\overline{w}\,\mathrm{d}x\,\mathrm{d}y\,\mathrm{d}z\,\mathrm{d}t,$$

which extends over the whole area of sickle, where (see (67) in §12)

$$\text{lor } \nu\overline{w} = \frac{\partial \nu w_1}{\partial x_1} + \frac{\partial \nu w_2}{\partial x_2} + \frac{\partial \nu w_3}{\partial x_3} + \frac{\partial \nu w_4}{\partial x_4}.$$

When the sickle shrunk to a spacetime point x, y, z, t we arrive at the differential equation

$$\text{lor } \nu\overline{w} = 0, \tag{6}$$

i.e. at *the continuity condition*

$$\frac{\partial \mu \mathfrak{w}_x}{\partial x} + \frac{\partial \mu \mathfrak{w}_y}{\partial y} + \frac{\partial \mu \mathfrak{w}_z}{\partial z} + \frac{\partial \mu}{\partial t} = 0.$$

We form also the integral which extends over the whole area of a spacetime sickle

$$N = \iiiint \nu \, dx \, dy \, dz \, dt. \tag{7}$$

We cut the sickle into thin the spacetime threads and cut each of these threads into small elements $d\tau$ of its proper time, which are however still large compared to the linear dimensions of the normal cross section of the thread. If we take the mass of such a thread to be $\nu \, dJ_n = dm$ and τ^0 and τ^1 are the proper time of the thread at the lower and the upper boundary of the sickle, respectively, the integral (7) can be interpreted as

$$\iint \nu \, dJ_n \, d\tau = \int \left(\tau^1 - \tau^0 \right) dm.$$

over all the threads in the sickle.

Now I take the spacetime lines within a spacetime sickle as material curves consisting of material points and think of them as being subjected to a continuous change of position within the sickle in the following way. All the curves are somehow *fixed at the end points of the lower and upper boundary of the sickle* and the individual material points on them always move *normal to the curves*. The whole process should be analytically represented by a parameter θ and the value $\theta = 0$ corresponds to the curves that represent the actual spacetime lines within the sickle. Such a process is called a *virtual displacement in the sickle*.

The point x, y, z, t in the sickle for $\theta = 0$ changes its position to $x + \delta x, y + \delta y, z + \delta z, t + \delta t$ for the value θ of parameter; the latter quantities are then functions of x, y, z, t, θ. If we consider again an infinitely thin

spacetime thread, the volume $\mathrm{d}J_n$ of its normal cross section at the initial
point x, y, z, t becomes the volume $\mathrm{d}J_n + \delta \, \mathrm{d}J_n$ of the normal cross section
at the corresponding point of the varied thread. When we take into account
the principle of conservation of mass and that the rest mass density at the
varied position is $\nu + \delta\nu$, where ν is the real rest mass density at x, y, z, t,
we can write

$$(\nu + \delta\nu)(\mathrm{d}J_n + \delta \, \mathrm{d}J_n) = \nu \, \mathrm{d}J_n = \mathrm{d}m. \tag{8}$$

On the basis of this result, the integral (7) taken over the area of the sickle,
varies due the virtual displacement as a some function $N + \delta N$ of θ, and we
will call this function $N + \delta N$ of θ the *mass effect* of the virtual displacement.

If we go by the notation with indices, we will have:

$$\mathrm{d}(x_h + \delta x_h) = \mathrm{d}x_h + \sum_k \frac{\partial \delta x_h}{\partial x_k} \mathrm{d}x_k + \frac{\partial \delta x_h}{\partial \theta} \mathrm{d}\theta \qquad \left(\begin{array}{l} k = 1, 2, 3, 4 \\ h = 1, 2, 3, 4 \end{array} \right). \tag{9}$$

Then on the basis of the comments already made above is immediately seen
that the value of $N + \delta N$ for the parameter θ will be

$$N + \delta N = \iiiint \nu \frac{\mathrm{d}(\tau + \delta\tau)}{\mathrm{d}\tau} \, \mathrm{d}x \, \mathrm{d}y \, \mathrm{d}z \, \mathrm{d}t \tag{10}$$

which is taken over the sickle, and where the magnitude $\mathrm{d}(\tau + \delta\tau)$ follows
from

$$\sqrt{-(\mathrm{d}x_1 + \mathrm{d}\delta x_1)^2 - (\mathrm{d}x_2 + \mathrm{d}\delta x_2)^2 - (\mathrm{d}x_3 + \mathrm{d}\delta x_3)^2 - (\mathrm{d}x_4 + \mathrm{d}\delta x_4)^2},$$

by taking account of (9), and

$$\mathrm{d}x_1 = w_1 \mathrm{d}\tau, \qquad \mathrm{d}x_2 = w_2 \mathrm{d}\tau, \qquad \mathrm{d}x_3 = w_3 \mathrm{d}\tau, \qquad \mathrm{d}x_4 = w_4 \mathrm{d}\tau, \qquad \mathrm{d}\theta = 0,$$

we have

$$\frac{\mathrm{d}(\tau + \delta\tau)}{\mathrm{d}\tau} = \sqrt{-\sum_k \left(w_h + \sum_k \frac{\partial \delta x_h}{\partial x_k} w_k \right)^2} \qquad \left(\begin{array}{l} k = 1, 2, 3, 4 \\ h = 1, 2, 3, 4 \end{array} \right). \tag{11}$$

Now we want to subject the value of the derivative

$$\left(\frac{\mathrm{d}(N+\delta N)}{\mathrm{d}\theta}\right)_{(\theta=0)} \tag{12}$$

to a transformation. Since each δx_h as a function of the arguments x, y, z, y, θ vanishes in general for $\theta = 0$, it is also generally $\frac{\partial \delta x_h}{\partial x_k} = 0$ for $\theta = 0$. If we now denote

$$\left(\frac{\partial \delta x_h}{\partial x_k}\right)_{\theta=0} = \xi_h \quad (h = 1, 2, 3, 4), \tag{13}$$

then taking account of (10) and (11) the expression (12) becomes

$$-\iiiint \nu \sum_h w_h \left(\frac{\partial \xi_h}{\partial x_1}w_1 + \frac{\partial \xi_h}{\partial x_2}w_2 + \frac{\partial \xi_h}{\partial x_3}w_3 + \frac{\partial \xi_h}{\partial x_4}w_4\right) \mathrm{d}x \ \mathrm{d}y \ \mathrm{d}z \ \mathrm{d}t.$$

For the system x_1, x_2, x_3, x_4 on the boundary of the sickle $\delta x_1, \delta x_2, \delta x_3, \delta x_4$ should vanish for every value of θ and therefore also $\xi_1, \xi_2, \xi_3, \xi_4$ are all zero. After that, the last integral transforms by integration by parts

$$\iiiint \sum_h \xi_h \left(\frac{\partial \nu w_h w_1}{\partial x_1} + \frac{\partial \nu w_h w_2}{\partial x_2} + \frac{\partial \nu w_h w_3}{\partial x_3} + \frac{\partial \nu w_h w_4}{\partial x_4}\right) \mathrm{d}x \ \mathrm{d}y \ \mathrm{d}z \ \mathrm{d}t.$$

The expression in the brackets

$$= w_h \sum_k \frac{\partial \nu w_k}{\partial x_k} + \nu \sum_k w_k \frac{\partial w_h}{\partial x_k}.$$

The first sum here vanishes according to the continuity condition (6), the second can be represented as

$$\frac{\partial w_h}{\partial x_1}\frac{\mathrm{d}x_1}{\mathrm{d}\tau} + \frac{\partial w_h}{\partial x_2}\frac{\mathrm{d}x_2}{\mathrm{d}\tau} + \frac{\partial w_h}{\partial x_3}\frac{\mathrm{d}x_3}{\mathrm{d}\tau} + \frac{\partial w_h}{\partial x_4}\frac{\mathrm{d}x_4}{\mathrm{d}\tau} = \frac{\mathrm{d}w_h}{\mathrm{d}\tau} = \frac{\mathrm{d}}{\mathrm{d}\tau}\left(\frac{\mathrm{d}x_h}{\mathrm{d}\tau}\right),$$

where $\frac{\mathrm{d}}{\mathrm{d}\tau}$ indicates differentiation in the direction of the spacetime line at a given point. Finally, *for the derivative(12) we have the expression*

$$\iiiint \nu \left(\frac{\mathrm{d}w_1}{\mathrm{d}\tau}\xi_1 + \frac{\mathrm{d}w_2}{\mathrm{d}\tau}\xi_2 + \frac{\mathrm{d}w_3}{\mathrm{d}\tau}\xi_3 + \frac{\mathrm{d}w_4}{\mathrm{d}\tau}\xi_4 \right) \mathrm{d}x \ \mathrm{d}y \ \mathrm{d}z \ \mathrm{d}t. \qquad (14)$$

For a virtual displacement in the sickle we have set the requirement that the material points should move normal to the curves produced from them, which means that for $\theta = 0$ the quantities ξ_h must satisfy the *condition*

$$w_1\xi_1 + w_2\xi_2 + w_3\xi_3 + w_4\xi_4 = 0. \qquad (15)$$

Consider now Maxwell's stresses in the electrodynamics of bodies at rest and, on the other hand, at our results in §12 and §13 then there is some *adjustment of the Hamilton principle* for continuously extended elastic media *to the relativity postulate.*

Let at each spacetime point (as in §13) a spacetime matrix of the II-nd kind be known

$$S = \begin{vmatrix} S_{11}, & S_{12}, & S_{13}, & S_{14} \\ S_{21}, & S_{22}, & S_{23}, & S_{24} \\ S_{31}, & S_{32}, & S_{33}, & S_{34} \\ S_{41}, & S_{42}, & S_{43}, & S_{44} \end{vmatrix} = \begin{vmatrix} X_x, & Y_x, & Z_x, & -iT_x \\ X_y, & Y_y, & Z_y, & -iT_y \\ X_z, & Y_z, & Z_z, & -iT_z \\ -iX_t, & -iY_t, & -iZ_t, & T_t \end{vmatrix}, \qquad (16)$$

where $X_x, Y_x, \cdots, Z_z, T_x, \cdots, X_t, \cdots, T_t$ are real quantities.

For a virtual displacement in a spacetime sickle with the previously used terms the value of the integral

$$W + \delta W = \iiiint \left(\sum_{h,k} S_{hk} \frac{\partial(x_k + \delta x_k)}{\partial x_h} \right) \mathrm{d}x \ \mathrm{d}y \ \mathrm{d}z \ \mathrm{d}t, \qquad (17)$$

which extends over the area of the sickle, may be called the *stress effect* in the virtual displacement.

The sum occurring here, given in more detail and written with real quantities, is

$$X_x + Y_y + Z_z + T_t$$

$$+X_x\frac{\partial \delta x}{\partial x} + X_y\frac{\partial \delta x}{\partial y} + \cdots + Z_z\frac{\partial \delta z}{\partial z}$$

$$-X_t\frac{\partial\delta x}{\partial t}-\cdots+T_x\frac{\partial\delta t}{\partial x}+\cdots+T_t\frac{\partial\delta t}{\partial t}.$$

We will now introduce the following *minimum principle for mechanics*:

For any bounded spacetime sickle the sum of the mass effect and the stress effect should always be an extremum for each virtual displacement in the sickle for the actually occurring configurations of the spacetime lines in the sickle.

The meaning of this statement is that, in the previously explained symbols, for every virtual displacement it should be

$$\left(\frac{\mathrm{d}(\delta N+\delta w)}{\mathrm{d}\theta}\right)_{\theta=0}=0. \tag{18}$$

By using the methods of the calculus of variations the four differential equations below follow immediately from the minimum principle under consideration after taking into account the condition (15) and using the transformation (14)

$$\nu\frac{\mathrm{d}w_h}{\mathrm{d}\tau}=K_h+\varkappa w_h \qquad (h=1,2,3,4), \tag{19}$$

where

$$K_h=\frac{\partial S_{1h}}{\partial x_1}+\frac{\partial S_{2h}}{\partial x_2}+\frac{\partial S_{3h}}{\partial x_3}+\frac{\partial S_{4h}}{\partial x_4} \tag{20}$$

are the components of the spacetime vector of the I-st kind $K=$ lor S, and $\varkappa$ is a factor whose determination has to be made on the basis of $w\overline{w}=-1$. By multiplying each equation of (19) with w_h, and subsequent summation over $h=1,2,3,4$ we find $\varkappa=K\overline{w}$ and obviously $K+(K\overline{w})w$ is a spacetime vector of the I-st kind, which is normal to w. If we write the components of this vector

$$X,\ Y,\ Z,\ iT,$$

we now come to the following *laws for the motion of matter*:

$$\nu\frac{\mathrm{d}}{\mathrm{d}\tau}\frac{\mathrm{d}x}{\mathrm{d}\tau} = X,$$

$$\nu\frac{\mathrm{d}}{\mathrm{d}\tau}\frac{\mathrm{d}y}{\mathrm{d}\tau} = Y,$$

$$\nu\frac{\mathrm{d}}{\mathrm{d}\tau}\frac{\mathrm{d}z}{\mathrm{d}\tau} = Z,$$ (21)

$$\nu\frac{\mathrm{d}}{\mathrm{d}\tau}\frac{\mathrm{d}t}{\mathrm{d}\tau} = T.$$

We have

$$\left(\frac{\mathrm{d}x}{\mathrm{d}\tau}\right)^2 + \left(\frac{\mathrm{d}y}{\mathrm{d}\tau}\right)^2 + \left(\frac{\mathrm{d}z}{\mathrm{d}\tau}\right)^2 = \left(\frac{\mathrm{d}t}{\mathrm{d}\tau}\right)^2 - 1,$$

and

$$X\frac{\mathrm{d}x}{\mathrm{d}\tau} + Y\frac{\mathrm{d}y}{\mathrm{d}\tau} + Z\frac{\mathrm{d}z}{\mathrm{d}\tau} = T\frac{\mathrm{d}t}{\mathrm{d}\tau},$$

and on the basis of these circumstances the fourth equation in (21) can be viewed as a consequence of the first three.

From (21) we derive further the laws for the motion of a *material point*, that is to say for the course of an infinitely thin spacetime thread.

Let x, y, z, t denote a point on the accepted main line in the thread. We form the equations (21) for the points of the *normal cross section* of the thread through x, y, z, t and integrate them, which is to multiply them by the volume element of the cross section, over the whole space of the normal cross section.

If the integrals on the right side of (21) are R_x, R_y, R_z, R_t and m is the constant mass of the thread we can write

$$m\frac{\mathrm{d}}{\mathrm{d}\tau}\frac{\mathrm{d}x}{\mathrm{d}\tau} = R_x,$$

$$m\frac{\mathrm{d}}{\mathrm{d}\tau}\frac{\mathrm{d}y}{\mathrm{d}\tau} = R_y,$$

$$m\frac{\mathrm{d}}{\mathrm{d}\tau}\frac{\mathrm{d}z}{\mathrm{d}\tau} = R_z,$$ (22)

$$m\frac{\mathrm{d}}{\mathrm{d}\tau}\frac{\mathrm{d}t}{\mathrm{d}\tau} = R_t.$$

Here again R with the components R_x, R_y, R_z, iR_t is a spacetime vector of the I-st kind, which is normal to the spacetime vector of the I-st kind w, the velocity of the material point, with the components

$$\frac{dx}{d\tau}, \frac{dy}{d\tau}, \frac{dz}{d\tau}, i\frac{dt}{d\tau}.$$

We will call this vector R the *moving force* of the material point.

If, however, instead of integrating the equations above over the normal cross section of the thread, we integrate them over the cross section that is normal to the $t-$axis and passes through x, y, z, t, the results is (see (4)) that the equations (22), multiplied with $\frac{d\tau}{dt}$, in particular the last equation which becomes

$$m\frac{d}{dt}\left(\frac{dt}{d\tau}\right) = \mathfrak{w}_x R_x \frac{d\tau}{dt} + \mathfrak{w}_y R_y \frac{d\tau}{dt} + \mathfrak{w}_z R_z \frac{d\tau}{dt}.$$

Now the right side can be interpreted as the *work done per unit time (Arbeitsleistung)* on the material point. This equation itself is then the *law of energy (Energiesatz)* for the motion of the material point, and the expression

$$m\left(\frac{dt}{d\tau} - 1\right) = m\left(\frac{1}{\sqrt{1 - \mathfrak{w}^2}} - 1\right) = m\left(\frac{1}{2}|\mathfrak{w}|^2 + \frac{3}{8}|\mathfrak{w}|^4 + \cdots\right)$$

will be seen as the kinetic energy of the material point.

Since always $dt > d\tau$ the quotient $\frac{dt - d\tau}{d\tau}$ could be called the advance (Vorgehen) of time in comparison with the proper time of the material point, and then we can say: The kinetic energy of a material point is the product of its mass and the advance of time in comparison with its proper time.

The quadruple of equations (22) again shows the required by the relativity postulate full symmetry in the x, y, z, it, *while to the fourth equation*, as we have already encountered an analogous case in electrodynamics, *a higher physical evidence (Evidenz) should be attributed.* Due to the requirement of symmetry along the lines of the fourth equation we can now set up the triple of the first three equations and in view of this fact, the following statement is justified: *When the relativity postulate is put at the head of mechanics the complete laws of motion follow solely from the law of energy.*

I would not omit to make even more plausible that it is not expected from the phenomena of gravitation to produce an opposition to the acceptance of the relativity postulate[12].

---

[12]In a completely different way, from the one I follow here, H. Poincaré tried to adapt

If $B^*(x^*, y^*, z^*, t^*)$ is a fixed spacetime point, the area of all those space-time points $B(x, y, z, t)$ for which

$$(x - x^*)^2 + (y - y^*)^2 + (z - z^*)^2 = (t - t^*)^2, \qquad t - t^* \geq 0 \qquad (23)$$

is called the *lightray structure (Strahlgebilde)* of the spacetime point B^*.

This structure intersects an arbitrary spacetime line always only in a single spacetime point B, which, on the one hand, follows from the convexity of the structure, and, on the other hand, from the circumstance that all directions of a spacetime line are only directions from B^* toward the concave side of the structure. We will call B^* the *point light source (Lichtpunkt)* of B.

If in the condition (23) the point $B(x, y, z, t)$ is regarded as fixed, whereas point $B^*(x^*, y^*, z^*, z^*)$ is thought as variable, then the same relation defines the structure consisting of all spacetime points B^*, which are point light sources of B, and analogously it turns out that on any spacetime line there is always only a single point B^*, which is the point light source of B.

There may now be a material point F of mass m, which in the presence of another material point of mass m^* experiences a moving force according to the following law. Imagine the spacetime threads of F and F^* with the main lines in them. Let us take an infinitely small element BC on the main line of F, further on the main line of F^*, B^* is the point light source of B and C^* is the point light source of C; then OA' is a radius vector of a point on the hyperboloidal fundamental structure (2) parallel to B^*C^*, and finally D^* is the intersection point of the line B^*C^* with the space which is normal to it and passes through B. *The moving force of the point mass F at the spacetime point B may now be a spacetime vector of the I-st kind, which is normal to BC, and which is additively composed of the vector*

$$mm^* \left(\frac{OA'}{B^*D^*} \right)^3 BD^* \qquad (24)$$

in the direction of BD^* and of a suitable vector in the direction of B^*C^*. Here $\frac{OA'}{B^*D^*}$ is the ratio of the two respective parallel vectors.

It is clear that this construction has a covariant character with respect to the Lorentz group.

Newton's law of attraction to the relativity postulate (Rendiconti del Circolo Matematico di Palermo, T. XXI (1906), p. 129).

We now ask how the spacetime thread of F behaves, if the material point F^* is in a uniform translational motion, which means that the main line of the thread of F^* is a straight line. We move the spacetime origin 0 to it, and by a Lorentz transformation we can regard this line as the t−axis. Now x, y, z, t represent the point B and τ^* is the proper time of the point B^*, calculated from O. Our construction here leads to the equations

$$\frac{\mathrm{d}^2 x}{\mathrm{d}\tau^2} = -\frac{m^* x}{(t - \tau^*)^3}, \qquad \frac{\mathrm{d}^2 y}{\mathrm{d}\tau^2} = -\frac{m^* y}{(t - \tau^*)^3}, \qquad \frac{\mathrm{d}^2 z}{\mathrm{d}\tau^2} = -\frac{m^* z}{(t - \tau^*)^3} \qquad (25)$$

and

$$\frac{\mathrm{d}^2 t}{\mathrm{d}\tau^2} = -\frac{m^*}{(t - \tau^*)^2} \frac{\mathrm{d}(t - \tau^*)}{\mathrm{d}t}, \qquad (26)$$

where

$$x^2 + y^2 + z^2 = (t - \tau^*)^2 \qquad (27)$$

and

$$\left(\frac{\mathrm{d}x}{\mathrm{d}\tau}\right)^2 + \left(\frac{\mathrm{d}y}{\mathrm{d}\tau}\right)^2 + \left(\frac{\mathrm{d}z}{\mathrm{d}\tau}\right)^2 = \left(\frac{\mathrm{d}t}{\mathrm{d}\tau}\right)^2 - 1. \qquad (28)$$

The three equations (25) in view of (27) are exactly as the equations for the motion of a material point according to Newton's laws for a fixed center of attraction, except that instead of the time t the proper time τ of the material point occurs. The fourth equation (26) then gives the relation between proper time and time for the material point.

Now be the orbit of the spatial point x, y, z for different τ is an ellipse with a semi-major axis a, eccentricity e, and let E be the eccentric anomaly, then T is the increase in proper time for a full revolution on the orbit, finally $nT = 2\pi$ with a suitable initial value of τ we have the Kepler equation

$$n\tau = E - e \sin E \qquad (29)$$

Changing the unit of time and again denoting the velocity of light by c, it follows from (28):

$$\left(\frac{dt}{d\tau}\right)^2 - 1 = \frac{m^*}{ac^2}\frac{1+e\cos E}{1-e\cos E} \tag{30}$$

Neglecting terms proportional to c^{-4} compared to 1, it follows that

$$n\,dt = n\,d\tau\left(1 + \frac{1}{2}\frac{m^*}{ac^2}\frac{1+e\cos E}{1-e\cos E}\right),$$

which together with the use of (29) is

$$nt + const = \left(1 + \frac{1}{2}\frac{m^*}{ac^2}\right)n\tau + \frac{m^*}{ac^2}\sin E. \tag{31}$$

The factor $\frac{m^*}{ac^2}$ here is the square of the ratio of a some average orbital velocity of F to the velocity of light. If m^* is the mass of the Sun and a is half the major axis of the Earth's, then this factor is 10^{-8}.

The law of attraction for masses according to the formulation just discussed and related to the relativity postulate would also imply *a propagation of gravitation with the velocity of light*. Given the smallness of the periodic term in (31), a decision should not be inferred from astronomical observations *against* such a law and the proposed modified mechanics and in favor of the Newtonian law of attraction of the Newtonian mechanics.

Space and Time

Gentlemen! The views of space and time which I want to present to you arose from the domain of experimental physics, and therein lies their strength. Their tendency is radical. From now onwards space by itself and time by itself will recede completely to become mere shadows and only a type of union of the two will still stand independently on its own.

I.

I want to show first how to move from the currently adopted mechanics through purely mathematical reasoning to modified ideas about space and time. The equations of Newtonian mechanics show a twofold invariance. First, their form is preserved when subjecting the specified spatial coordinate system to *any change of position*; second, when it changes its state of motion, namely when any *uniform translation* is impressed upon it; also, the zero point of time plays no role. When one feels ready for the axioms of mechanics, one is accustomed to regard the axioms of geometry as settled and probably for this reason those two invariances are rarely mentioned in the same breath. Each of them represents a certain group of transformations for the differential equations of mechanics. The existence of the first group can be seen as reflecting a fundamental characteristic of space. One always tends to treat the second group with disdain in order to unburden one's mind that one can never determine from physical phenomena whether space, which is assumed to be at rest, may not after all be in uniform translation. Thus these two groups lead completely separate lives side by side. Their entirely heterogeneous character may have discouraged any intention to compose them. But it is the composed complete group as a whole that gives us to think.

We will attempt to visualize the situation graphically. Let x, y, z be orthogonal coordinates for space and let t denote time. The objects of our perception are always connected to places and times. No one has noticed a place other than at a time and a time other than at a place. However I still respect the dogma that space and time each have an independent meaning. I will call a point in space at a given time, i.e. a system of values x, y, z, t a *worldpoint*. The manifold of all possible systems of values x, y, z, t will be called the *world*. With a hardy piece of chalk I can draw four world axes on the blackboard. Even *one* drawn axis consists of nothing but vibrating molecules and also makes the journey with the Earth in the Universe, which already requires sufficient abstraction; the somewhat greater abstraction associated with the number 4 does not hurt the mathematician. To never let a yawning emptiness, let us imagine that everywhere and at any time something perceivable exists. In order not to say matter or electricity I will use the word substance for that thing. We focus our attention on the substantial point existing at the worldpoint x, y, z, t and imagine that we can recognize this substantial point at any other time. A time element dt may correspond to the changes dx, dy, dz of the spatial coordinates of this substantial point. We then get an image, so to say, of the eternal course of life of the substantial point, a curve in the world, a *worldline*, whose points can be clearly related to the parameter t from $-\infty$ to $+\infty$. The whole world presents itself as resolved into such worldlines, and I want to say in advance, that in my understanding the laws of physics can find their most complete expression as interrelations between these worldlines.

Through the concepts of space and time the x, y, z-manifold $t = 0$ and its two sides $t > 0$ and $t < 0$ fall apart. If for simplicity we hold the chosen origin of space and time fixed, then the first mentioned group of mechanics means that we can subject the x, y, z-axes at $t = 0$ to an arbitrary rotation about the origin corresponding to the homogeneous linear transformations of the expression

$$x^2 + y^2 + z^2.$$

The second group, however, indicates that, also without altering the expressions of the laws of mechanics, we may replace

$$x, y, z, t \quad \text{by} \quad x - \alpha t, \ y - \beta t, \ z - \gamma t, \ t,$$

where α, β, γ are any constants. The time axis can then be given a completely arbitrary direction in the upper half of the world $t > 0$. What has now the requirement of orthogonality in space to do with this complete freedom of choice of the direction of the time axis upwards?

To establish the connection we take a positive parameter c and look at the structure

$$c^2t^2 - x^2 - y^2 - z^2 = 1.$$

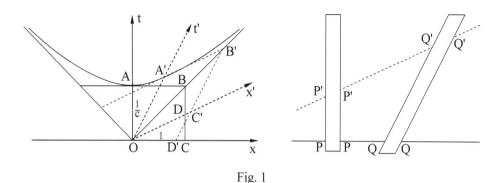

Fig. 1

It consists of two sheets separated by $t = 0$ by analogy with a two-sheeted hyperboloid. We consider the sheet in the region $t > 0$ and we will now take those homogeneous linear transformations of x, y, z, t in four new variables x', y', z', t' so that the expression of this sheet in the new variables has the same form. Obviously, the rotations of space about the origin belong to these transformations. A full understanding of the rest of those transformations can be obtained by considering such among them for which y and z remain unchanged. We draw (Fig. 1) the intersection of that sheet with the plane of the x- and the t-axis, i.e. the upper branch of the hyperbola $c^2t^2 - x^2 = 1$ with its asymptotes. Further we draw from the origin O an arbitrary radius vector OA' of this branch of the hyperbola; then we add the tangent to the hyperbola at A' to intersects the right asymptote at B'; from $OA'B'$ we complete the parallelogram $OA'B'C'$; finally, as we will need it later, we extend $B'C'$ so that it intersects the x-axis at D'. If we now regard OC' and OA' as axes for new coordinates x', t', with the scale units $OC' = 1$, $OA' = 1/c$, then that branch of the hyperbola again obtains the expression $ct'^2 - x'^2 = 1$, $t' > 0$, and the transition from x, y, z, t to x', y', z', t' is one of the transformations in question. These transformations plus the arbitrary displacements of the origin of space and time constitute a group of transformations which still depends on the parameter c and which I will call G_c.

If we now increase c to infinity, so $1/c$ converges to zero, it is clear from the figure that the branch of the hyperbola leans more and more towards the x-axis, that the angle between the asymptotes becomes greater, and in the limit that special transformation converts to one where the t'-axis may be in any upward direction and x' approaches x ever more closely. By taking this into account it becomes clear that the group G_c in the limit $c = \infty$, that is the group G_∞, is exactly the complete group which is associated with the Newtonian mechanics. In this situation, and since G_c is mathematically more understandable than G_∞, there could have probably been a mathematician with a free imagination who could have come up with the idea that at the end natural phenomena do not actually possess an invariance with the group G_∞, but rather with a group G_c with a certain finite c, which is *extremely great* only in the ordinary units of measurement. Such an insight would have been an extraordinary triumph for pure mathematics. Now mathematics expressed only staircase wit here, but it has the satisfaction that, due to its happy antecedents with their senses sharpened by their free and penetrating imagination, it can grasp the profound consequences of such remodelling of our view of nature.

I want to make it quite clear what the value of c will be with which we will be finally dealing. c is the *velocity of the propagation of light in empty space.* To speak neither of space nor of emptiness, we can identify this magnitude with the ratio of the electromagnetic to the electrostatic unit of the quantity of electricity.

The existence of the invariance of the laws of nature with respect to the group G_c would now be stated as follows:

From the entirety of natural phenomena, through successively enhanced approximations, it is possible to deduce more precisely a reference system x, y, z, t, space and time, by means of which these phenomena can be then represented according to certain laws. But this reference system is by no means unambiguously determined by the phenomena. *One can still change the reference system according to the transformations of the above group G_c arbitrarily without changing the expression of the laws of nature in the process.*

For example, according to the figure depicted above one can call t' time, but then must necessarily, in connection with this, define space by the manifold of three parameters x', y, z in which the laws of physics would then have exactly the same expressions by means of x', y, z, t' as by means of x, y, z, t. Hereafter we would then have in the world no more *the* space, but an infinite number of spaces analogously as there is an infinite number of planes in three-dimensional space. Three-dimensional geometry becomes a chapter in four-dimensional physics. You see why I said at the beginning that space and time will recede completely to become mere shadows and only a world in itself will exist.

II.

Now the question is, what circumstances force us to the changed view of space and time, does it actually never contradict the phenomena, and finally, does it provide advantages for the description of the phenomena?

Before we discuss these questions, an important remark is necessary. Having individualized space and time in some way, a straight worldline parallel to the t-axis corresponds to a stationary substantial point, a straight line inclined to the t-axis corresponds to a uniformly moving substantial point, a somewhat curved worldline corresponds to a non-uniformly moving substantial point. If at any worldpoint x, y, z, t there is a worldline passing through it and we find it parallel to any radius vector OA' of the previously mentioned hyperboloidal sheet, we may introduce OA' as a new time axis, and with the thus given new concepts of space and time, the substance at the worldpoint in question appears to be at rest. We now want to introduce this fundamental axiom:

With appropriate setting of space and time the substance existing at any worldpoint can always be regarded as being at rest.

This axiom means that at every worldpoint[1] the expression

$$c^2 \mathrm{d}t^2 - \mathrm{d}x^2 - \mathrm{d}y^2 - \mathrm{d}z^2$$

is always positive, which is equivalent to saying that any velocity v is always smaller than c. Then c would be an upper limit for all substantial velocities and that is precisely the deeper meaning of the quantity c. In this understanding the axiom is at first glance slightly displeasing. It should be noted, however, that a modified mechanics, in which the square root of that second order differential expression enters, is now gaining ground, so that cases with superluminal velocity will play only such a role as that of figures with imaginary coordinates in geometry.

The *impulse* and true motivation for *accepting the group G_c* came from noticing that the differential equation for the propagation of light waves in the empty space possesses that group G_c[2]. On the other hand, the concept of a rigid body has meaning only in a mechanics with the group G_∞. If one has optics with G_c, and if, on the other hand, there were rigid bodies, it is easy to see that *one* t-direction would be distinguished by the two hyperboloidal sheets corresponding to G_c and G_∞, and would have the further consequence that one would be able, by using appropriate rigid optical instruments in

---

[1] *Editor's note:* Minkowski means at every worldpoint along the worldline of the substance.

[2] An important application of this fact can already be found in W. Voigt, Göttinger Nachrichten, 1887, S. 41.

the laboratory, to detect a change of phenomena at various orientations
with respect to the direction of the Earth's motion. All efforts directed
towards this goal, especially a famous interference experiment of Michelson
had, however, a negative result. To obtain an explanation, H. A. Lorentz
made a hypothesis, whose success lies precisely in the invariance of optics
with respect to the group G_c. According to Lorentz every body moving at a
velocity v must experience a reduction in the direction of its motion namely
in the ratio

$$1 : \sqrt{1 - \frac{v^2}{c^2}}.$$

This hypothesis sounds extremely fantastical. Because the contraction is not
to be thought of as a consequence of resistances in the ether, but merely as
a gift from above, as an accompanying circumstance of the fact of motion.

I now want to show on our figure that the Lorentzian hypothesis is completely equivalent to the new concept of space and time, which makes it
much easier to understand. If for simplicity we ignore y and z and think
of a world of one spatial dimension, then two strips, one upright parallel
to the t-axis and the other inclined to the t-axis (see Fig. 1), are images
for the progression in time of a body at rest and a body moving uniformly,
where each preserves a constant spatial dimension. OA' is parallel to the
second strip, so we can introduce t' as time and x' as a space coordinate
and then it appears that the second body is at rest, whereas the first –
in uniform motion. We now assume that the first body has length l when
considered at rest, that is, the cross section PP of the first strip and the
x-axis is equal to $l \cdot OC$, where OC is the measuring unit on the x-axis,
and, on the other hand, that the second body has the same length l when
regarded at rest; then the latter means that the cross-section of the second
strip *measured parallel to the x'-axis* is $Q'Q' = l \cdot OC'$. We have now in
these two bodies images of two *equal* Lorentz electrons, one stationary and
one uniformly moving. But if we go back to the original coordinates x, t, we
should take as the dimension of the second electron the cross section QQ of
its associated strip *parallel to the x-axis*. Now as $Q'Q' = l \cdot OC'$, it is obvious
that $QQ = l \cdot OD'$. If dx/dt for the second strip is $= v$, an easy calculation
gives $OD' = OC \cdot \sqrt{1 - \frac{v^2}{c^2}}$, therefore also $PP : QQ = 1 : \sqrt{1 - \frac{v^2}{c^2}}$. This is
the meaning of the Lorentzian hypothesis of the contraction of electrons in
motion. Regarding, on the other hand, the second electron as being at rest,
that is, adopting the reference system x', t', the length of the first electron
will be the cross section $P'P'$ of its strip parallel to OC', and we would find

the first electron shortened with respect to the second in exactly the same proportion; from the figure we also see that

$$P'P' : Q'Q' = OD : OC' = OD' : OC = QQ : PP.$$

Lorentz called t', which is a combination of x and t, *local time* of the uniformly moving electron, and associated a physical construction with this concept for a better understanding of the contraction hypothesis. However, it is to the credit of A. Einstein[3] who first realized clearly that the time of one of the electrons is as good as that of the other, i.e. that t and t' should be treated equally. With this, time was deposed from its status as a concept unambiguously determined by the phenomena. The concept of space was shaken neither by Einstein nor by Lorentz, maybe because in the above-mentioned special transformation, where the plane of x', t' coincides with the plane x, t, an interpretation is possible as if the x-axis of space preserved its position. To step over the concept of space in such a way is an instance of what can be achieved only due to the audacity of mathematical culture. After this further step, which is indispensable for the true understanding of the group G_c, I think the word *relativity postulate* used for the requirement of invariance under the group G_c is very feeble. Since the meaning of the postulate is that through the phenomena only the four-dimensional world in space and time is given, but the projection in space and in time can still be made with certain freedom, I want to give this affirmation rather the name *the postulate of the absolute world* (or shortly the world postulate).

III.

Through the world postulate an identical treatment of the four identifying quantities x, y, z, t becomes possible. I want to explain now how, as a result of this, we gain more understanding of the forms under which the laws of physics present themselves. Especially the concept of *acceleration* acquires a sharply prominent character.

I will use a geometric way of expression, which presents itself immediately when one implicitly ignores z in the triple x, y, z. An arbitrary worldpoint O can be taken as the origin of space-time. The *cone*

$$c^2t^2 - x^2 - y^2 - z^2 = 0$$

with O as the apex (Fig. 2) consists of two parts, one with values $t < 0$, the other one with values $t > 0$.

---

[3] A. Einstein, Annalen der Physik 17 (1905), S. 891; Jahrbuch der Radioaktivität und ELektronik 4 (1907), S. 411.

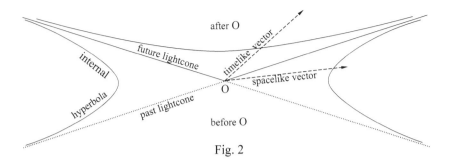

Fig. 2

The first, the *past lightcone of O*, consists, we can say, of all worldpoints which "send light to O", the second, the *future lightcone of O*, consists of all worldpoints which "receive light from O"[4]. The area bounded solely by the past lightcone may be called *before O*, whereas the area bounded solely by the future lightcone – *after O*. Situated after O is the already considered hyperboloidal sheet

$$F = c^2 t^2 - x^2 - y^2 - z^2 = 1, \ t > 0$$

The area *between the cones* is filled with the one-sheeted hyperboloidal structures

$$-F = x^2 + y^2 + z^2 - c^2 t^2 = k^2$$

for all constant positive values of k^2. Essential for us are the hyperbolas with O as the center, located on the latter structures. The individual branches of these hyperbolas may be briefly called *internal hyperbolas with center O*. Such a hyperbola would be thought of as the worldline of a substantive point, which represents its motion that increases asymptotically to the velocity of light c for $t = -\infty$ and $t = +\infty$.

If we now call, by analogy with vectors in space, a directed line in the manifold x, y, z, t a vector, we have to distinguish between the *timelike* vectors with directions from O to the sheet $+F = 1, t > 0$, and the *spacelike* vectors with directions from O to $-F = 1$. The time axis can be parallel to any vector of the first kind. Every worldpoint between the future lightcone

---

[4]*Editor's and translator's note:* I decided to translate the words *Vorkegel* and *Nachkegel* as *past lightcone* and *future lightcone*, respectively, for two reasons. First, this translation reflects the essence of Minkowski's idea – (i) all worldpoints on the past lightcone "send *light* to O", which means that they all can influence O and therefore lie in the past of O; (ii) all worldpoints on the future lightcone "receive *light* from O", which means that they all can be influenced by O and therefore lie in the *future* of O. Second, the terms *past lightcone* and *future lightcone* are now widely accepted in spacetime physics.

and the past lightcone of O can be regarded, by a choice of the reference system, as *simultaneous* with O as well as *earlier* than O or *later* than O. Each worldpoint within the past lightcone of O is necessarily always earlier than O, each worldpoint within the future lightcone is necessarily always later than O. The transition to the limit $c = \infty$ would correspond to a complete folding of the wedge-shaped section between the cones into the flat manifold $t = 0$. In the figures this section is intentionally made with different widths.

We decompose any vector, such as that from O to x, y, z, t into four *components* x, y, z, t. If the directions of two vectors are, respectively, that of a radius vector OR from O to one of the surfaces $\mp F = 1$, and that of a tangent RS at the point R on the same surface, the vectors are called *normal* to each other. Accordingly,

$$c^2 t t_1 - x x_1 - y y_1 - z z_1 = 0$$

is the condition for the vectors with components x, y, z, t and x_1, y_1, z_1, t_1 to be normal to each other.

The *measuring units* for the *magnitudes* of vectors in different directions may be fixed by assigning to a spacelike vector from O to $-F = 1$ always the magnitude 1, and to a timelike vector from O to $+F = 1, t > 0$ always the magnitude $1/c$.

Let us now imagine a worldpoint $P(x, y, z, t)$ through which the worldline of a substantial point is passing, then the magnitude of the timelike vector dx, dy, dz, dt along the line will be

$$d\tau = \frac{1}{c}\sqrt{c^2 dt^2 - dx^2 - dy^2 - dz^2}.$$

The integral $\int d\tau = \tau$ of this magnitude, taken along the worldline from any fixed starting point P_0 to the variable end point P, we call the *proper time* of the substantial point at P. On the worldline we consider x, y, z, t, i.e. the components of the vector OP, as functions of the proper time τ; denote their first derivatives with respect to τ by $\dot{x}, \dot{y}, \dot{z}, \dot{t}$; their second derivatives with respect to τ by $\ddot{x}, \ddot{y}, \ddot{z}, \ddot{t}$, and call the corresponding vectors, the derivative of the vector OP with respect to τ the *velocity vector at P* and the derivative of the velocity vector with respect to τ the *acceleration vector at P*. As

$$c^2 \dot{t}^2 - \dot{x}^2 - \dot{y}^2 - \dot{z}^2 = c^2$$

it follows that

$$c^2 \dot{t}\ddot{t} - \dot{x}\ddot{x} - \dot{y}\ddot{y} - \dot{z}\ddot{z} = 0,$$

i.e. the velocity vector is the timelike vector of magnitude 1 in the direction of the worldline at P, and the acceleration vector at P is normal to the velocity vector at P, so it is certainly a spacelike vector.

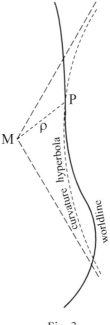

Fig. 3

Now there is, as is easily seen, a specific branch of the hyperbola, which has three infinitely adjacent points in common with the worldline at P, and whose asymptotes are generators of a past lightcone and a future lightcone (see Fig. 3). This branch of the hyperbola will be called the *curvature hyperbola* at P. If M is the center of this hyperbola, we have here an internal hyperbola with center M. Let ρ be the magnitude of the vector MP, *so we recognize the acceleration vector at P as the vector in the direction MP of magnitude c^2/ρ.*

If $\ddot{x}, \ddot{y}, \ddot{z}, \ddot{t}$ are all zero, the curvature hyperbola reduces to the straight line touching the worldline at P, and we should set $\rho = \infty$.

IV.

To demonstrate that the adoption of the group G_c for the laws of physics never leads to a contradiction, it is inevitable to undertake a revision of all physics based on the assumption of this group. This revision has been done successfully to some extent for questions of thermodynamics and heat radiation[5], for the electromagnetic processes, and finally, with the retention of the concept of mass, for mechanics.[6]

For the latter domain, the question that should be raised above all is: When a force with the spatial components X, Y, Z acts at a worldpoint $P(x, y, z, t)$, where the velocity vector is $\dot{x}, \dot{y}, \dot{z}, \dot{t}$, as what force this force should be interpreted for any change of the reference system? Now there exist some proven approaches to the ponderomotive force in the electromagnetic field in cases where the group G_c is undoubtedly permissible. These approaches lead to the simple rule: *When the reference system is changed, the given force transforms into a force in the new space coordinates in such a way that the corresponding vector with the components*

$$\dot{t}X, \ \dot{t}Y, \ \dot{t}Z, \ \dot{t}T$$

remains unchanged, and where

$$T = \frac{1}{c^2}\left(\frac{\dot{x}}{\dot{t}}X + \frac{\dot{y}}{\dot{t}}Y + \frac{\dot{z}}{\dot{t}}Z\right)$$

is the work done by the force at the worldpoint divided by c^2. This vector is always normal to the velocity vector at P. Such a force vector, representing a force at P, will be called a *motive force vector* at P.

Now let the worldline passing through P represent a substantial point with constant *mechanical mass* m. The multiplied by m velocity vector at P will be called the *momentum vector at* P, and the multiplied by m acceleration vector at P will be called the *force vector of the motion at* P. According to these definitions, the law of motion for a point mass with a given force vector is:[7]

The force vector of the motion is equal to the motive force vector.

This assertion summarizes four equations for the components for the four axes, wherein the fourth can be regarded as a consequence of the first three because both vectors are from the start normal to the velocity vector. According to the above meaning of T, the fourth equation is undoubtedly

---

[5]M. Planck, "Zur Dynamik bewegter Systeme," Sitzungsberichte der k. preußischen Akademie der Wissenschaften zu Berlin, 1907, S. 542 (auch Annalen der Physik, Bd. 26, 1908, S. 1).

[6]H. Minkowski, "Die Grundgleichungen für die elektromagnetischen Vorgänge in bewegten Körpern", Nachrichten der k. Gesellschaft der Wissenschaft zu Göttingen, mathematisch-physikalische Klasse, 1908, S. 53 und Mathematische Annalen, Bd. 68, 1910, S. 527

[7]H. Minkowski, loc. cit., p. 107. Cf. also M. Planck, Verhandlungen der Physikalischen Gesellschaft, Bd. 4, 1906, S. 136.

the law of conservation of energy. The *kinetic energy* of the point mass is defined as the *component of the momentum vector along the t-axis multiplied by* c^2. The expression for this is

$$mc^2 \frac{dt}{d\tau} = \frac{mc^2}{\sqrt{1 - \frac{v^2}{c^2}}},$$

which is, the expression $\frac{1}{2}mv^2$ of Newtonian mechanics after the subtraction of the additive constant term mc^2 and neglecting magnitudes of the order $1/c^2$. The *dependence of the energy on the reference system* is manifested very clearly here. But since the t-axis can be placed in the direction of each timelike vector, then, on the other hand, the law of conservation of energy, formed for every possible reference system, already contains the whole system of the equations of motion. In the discussed limiting case $c = \infty$, this fact will retain its importance for the axiomatic structure of Newtonian mechanics and in this sense has been already noticed by J. R. Schütz[8]

From the beginning we can determine the ratio of the units of length and time in such a way that the natural limit of velocity becomes $c = 1$. If we introduce $\sqrt{-1}t = s$ instead of t, then the quadratic differential expression

$$d\tau^2 = -dx^2 - dy^2 - dz^2 - ds^2$$

becomes completely symmetric in x, y, z, s and this symmetry is carried over to any law that does not contradict the world postulate. Thus the essence of this postulate can be expressed mathematically very concisely in the mystical formula:

$$3 \cdot 10^5 \ km = \sqrt{-1} \ seconds.$$

V.

The advantages resulting from the world postulate may most strikingly be proved by indicating the effects from *an arbitrarily moving point charge* according to the Maxwell-Lorentz theory. Let us imagine the worldline of such a pointlike electron with charge e, and take on it the proper time τ from any initial point. To determine the field induced by the electron at any worldpoint P_1 we construct the past lightcone corresponding to P_1 (Fig. 4). It intersects the infinite worldline of the electron obviously at a single point P

---

[8]J. R. Schütz, "Das Prinzip der absoluten Erhaltung der Energie", Nachrichten der k. Gesellschaft der Wissenschaften zu Göttingen, mathematisch-physikalische Klasse, 1897, S. 110.

because the tangents to every point on the worldline are all timelike vectors. At P we draw the tangent to the worldline and through P_1 construct the normal P_1Q to this tangent. Let the magnitude of P_1Q be r. According to the definition of a past lightcone the magnitude of PQ should be r/c. *Now the vector of magnitude e/r in the direction PQ represents through its components along the x-, y-, z-axes, the vector potential multiplied by c, and through the component along the t-axis, the scalar potential of the field produced by e at the worldpoint P_1.* This is the essence of the elementary laws formulated by A. Liénard and E. Wiechert.[9]

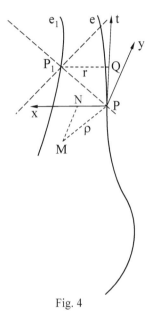

Fig. 4

Then it emerges in the description itself of the field caused by the electron that the division of the field into electric and magnetic forces is a relative one with respect to the specified time axis; most clearly the two forces considered together can be described in some, though not complete, analogy with the wrench in mechanics. I now want to describe *the ponderomotive action of an arbitrarily moving point charge exerted on another arbitrarily moving point charge.* Let us imagine that the worldline of a second pointlike electron of charge e_1 goes through the worldpoint P_1. We define P, Q, r as before, then

---

[9]A. Liénard, "Champ électrique et magnétique produit par une charge concentré en un point et animée d'un mouvement quelconque", L'Éclairage électrique, T. 16, 1898, pp. 5, 53, 106; E. Wiechert, "Elektrodynamische Elementargesetze", Archives Néerlandaiaes des Sciences exactes et naturelles (2), T. 5, 1900, S. 549.

construct (Fig. 4) the center M of the curvature hyperbola at P, and finally the normal MN from M to an imagined straight line from P parallel to QP_1. We now fix a reference system with its origin at P in the following way: the t-axis in the direction of PQ, the x-axis in the direction of QP_1, the y-axis in the direction of MN, and lastly the direction of the z-axis is determined as being normal to the t-, x-, y-axes. Let the acceleration vector at P be $\ddot{x}, \ddot{y}, \ddot{z}, \ddot{t}$, the velocity vector at P_1 be $\dot{x}_1, \dot{y}_1, \dot{z}_1, \dot{t}_1$. *Now the motive force vector exerted by the first arbitrarily moving electron e on the second arbitrarily moving electron e_1 at P_1 will be*

$$-ee_1(\dot{t}_1 - \frac{\dot{x}_1}{c})\mathfrak{K}$$

where for the components $\mathfrak{K}_x, \mathfrak{K}_y, \mathfrak{K}_z, \mathfrak{K}_t$ of the vector $\mathfrak{K}$ three relations exist:

$$c\mathfrak{K}_t - \mathfrak{K}_x = \frac{1}{r^2}, \quad \mathfrak{K}_y = \frac{\ddot{y}}{c^2 r}, \quad \mathfrak{K}_z = 0$$

and fourthly this vector $\mathfrak{K}$ is normal to the velocity vector at P_1, and this circumstance alone makes it dependent on the latter velocity vector.

If we compare this assertion with the previous formulations[10] of the same elementary law of the ponderomotive action of moving point charges on one another, we are compelled to admit that the relations considered here reveal their inner being in full simplicity only in four dimensions, whereas on a three dimensional space, forced upon us from the beginning, they cast only a very tangled projection.

In mechanics reformed in accordance with the world postulate, the disturbing disharmony between Newtonian mechanics and the modern electrodynamics disappears by itself. In addition, I want to touch on the status of the *Newtonian law of attraction* with respect to this postulate. I will consider two point masses m, m_1, represented by their worldlines, and that m exerts a motive force vector on m_1 exactly as in the case of electrons, except that instead of $-ee_1 +mm_1$ should be used. We can now specifically consider the case when the acceleration vector of m is constantly zero, then we may choose t in such a way that m is regarded as at rest, and assume that only m_1 move under the motive force vector which originates from m. If we now modify this specified vector by adding the factor $\dot{t}^{-1} = \sqrt{1 - \frac{v^2}{c^2}}$, which up to magnitudes of the order $1/c^2$ is equal to 1, it can be seen[11] that for the

---

[10]K. Schwazschild, Nachrichten der k. Gesellschaft der Wissenschaften zu Göttinger, mathematisch-physikalische Klasse, 1903, S. 132; H. A. Lorentz, Enzyklopädie der mathematischen Wissenschaften, V, Art. 14, S. 199.

[11]H. Minkowski, loc. cit., p. 110.

positions x_1, y_1, z_1 of m_1 and their progression in time, we arrive exactly at Kepler's laws, except that instead of the times t_1 the proper times τ_1 of m_1 should be used. On the basis of this simple remark we can then see that the proposed law of attraction associated with the new mechanics is no less well suited to explain the astronomical observations than the Newtonian law of attraction associated with the Newtonian mechanics.

The fundamental equations for the electromagnetic processes in ponderable bodies are entirely in accordance with the world postulate. Actually, as I will show elsewhere, there is no need to abandon the derivation of these equations which is based on ideas of the electron theory as taught by Lorentz.

The validity without exception of the world postulate is, I would think, the true core of an electromagnetic world view which, as Lorentz found it and Einstein further unveiled it, lies downright and completely exposed before us as clear as daylight. With the development of the mathematical consequences of this postulate, sufficient findings of its experimental validity will be arrived at so that even those to whom it seems unsympathetic or painful to abandon the prevailing views become reconciled through the thought of a pre-stabilized harmony between mathematics and physics.

Printed in Great Britain
by Amazon